U.S. Department
of Transportation

Federal Highway
Administration

FHWA-EP-00-005
DOT-VNTSC-FHWA-00-01

FHWA HIGHWAY NOISE BARRIER DESIGN HANDBOOK

Final Report
February 2000

REPRODUCED BY: **NTIS.**
U.S. Department of Commerce
National Technical Information Service
Springfield, Virginia 22161

Prepared for

U.S. Department of Transportation
Federal Highway Administration
Office of Natural Environment
Washington, D.C. 20590

Prepared by

U.S. Department of Transportation
Research and Special Programs Administration
John A. Volpe National Transportation Systems Center
Acoustics Facility, DTS-34
Cambridge, MA 02142-1093

NOTICE

This document is disseminated under the sponsorship of the Department
of Transportation in the interest of information exchange. The United
States Government assumes no liability for its contents or use
thereof. This report does not constitute a standard, specification,
or regulation.

The United States Government does not endorse products or
manufacturers. Trade or manufacturers' names appear herein solely
because they are considered essential to the object of this document.

PROTECTED UNDER INTERNATIONAL COPYRIGHT
ALL RIGHTS RESERVED
NATIONAL TECHNICAL INFORMATION SERVICE
U.S. DEPARTMENT OF COMMERCE

Reproduced from
best available copy.

GENERAL DISCLAIMER

This document may have problems that one or more of the following disclaimer statements refer to:

- This document has been reproduced from the best copy furnished by the sponsoring agency. It is being released in the interest of making available as much information as possible.

- This document may contain data which exceeds the sheet parameters. It was furnished in this condition by the sponsoring agency and is the best copy available.

- This document may contain tone-on-tone or color graphs, charts and/or pictures which have been reproduced in black and white.

- The document is paginated as submitted by the original source.

- Portions of this document are not fully legible due to the historical nature of some of the material. However, it is the best reproduction available from the original submission.

METRIC/ENGLISH CONVERSION FACTORS

ENGLISH TO METRIC

LENGTH (APPROXIMATE)

1 inch (in) = 2.5 centimeters (cm)
1 foot (ft) = 30 centimeters (cm)
1 yard (yd) = 0.9 meter (m)
1 mile (mi) = 1.6 kilometers (km)

AREA (APPROXIMATE)

1 square inch (sq in, in^2) = 6.5 square centimeters (cm^2)
1 square foot (sq ft, ft^2) = 0.09 square meter (m^2)
1 square yard (sq yd, yd^2) = 0.8 square meter (m^2)
1 square mile (sq mi, mi^2) = 2.6 square kilometers (km^2)
1 acre = 0.4 hectare (ha) = 4,000 square meters (m^2)

MASS - WEIGHT (APPROXIMATE)

1 ounce (oz) = 28 grams (gm)
1 pound (lb) = .45 kilogram (kg)
1 short ton = 2,000 pounds (lb) = 0.9 tonne (t)

VOLUME (APPROXIMATE)

1 teaspoon (tsp) = 5 milliliters (ml)
1 tablespoon (tbsp) = 15 milliliters (ml)
1 fluid ounce (fl oz) = 30 milliliters (ml)
1 cup (c) = 0.24 liter (l)
1 pint (pt) = 0.47 liter (l)
1 quart (qt) = 0.96 liter (l)
1 gallon (gal) = 3.8 liters (l)
1 cubic foot (cu ft, ft^3) = 0.03 cubic meter (m^3)
1 cubic yard (cu yd, yd^3) = 0.76 cubic meter (m^3)

TEMPERATURE (EXACT)

$°C = 5/9(°F - 32)$

METRIC TO ENGLISH

LENGTH (APPROXIMATE)

1 millimeter (mm) = 0.04 inch (in)
1 centimeter (cm) = 0.4 inch (in)
1 meter (m) = 3.3 feet (ft)
1 meter (m) = 1.1 yards (yd)
1 kilometer (km) = 0.6 mile (mi)

AREA (APPROXIMATE)

1 square centimeter (cm^2) = 0.16 square inch (sq in, in^2)
1 square meter (m^2) = 1.2 square yards (sq yd, yd^2)
1 square kilometer (km^2) = 0.4 square mile (sq mi, mi^2)
10,000 square meters (m^2) = 1 hectare (ha) = 2.5 acres

MASS - WEIGHT (APPROXIMATE)

1 gram (gm) = 0.036 ounce (oz)
1 kilogram (kg) = 2.2 pounds (lb)
1 tonne (t) = 1,000 kilograms (kg) = 1.1 short tons

VOLUME (APPROXIMATE)

1 milliliter (ml) = 0.03 fluid ounce (fl oz)
1 liter (l) = 2.1 pints (pt)
1 liter (l) = 1.06 quarts (qt)
1 liter (l) = 0.26 gallon (gal)
1 cubic meter (m^3) = 36 cubic feet (cu ft, ft^3)
1 cubic meter (m^3) = 1.3 cubic yards (cu yd, yd^3)

TEMPERATURE (EXACT)

$°F = 9/5(°C) + 32$

QUICK INCH-CENTIMETER LENGTH CONVERSION

INCHES	0		1		2		3		4		5
CENTIMETERS	0	1	2	3	4	5	6	7	8	9	10	11	12	13

QUICK FAHRENHEIT-CELSIUS TEMPERATURE CONVERSION

°F	-40°	-22°	-4°	14°	32°	50°	68°	86°	104°	122°	140°	158°	176°	194°	212°
°C	-40°	-30°	-20°	-10°	0°	10°	20°	30°	40°	50°	60°	70°	80°	90°	100°

For more exact and or other conversion factors, see NIST Miscellaneous Publication 286, Units of Weights and Measures. Price $2.50. SD Catalog No. C13 10286.

PREFACE

The U.S. Department of Transportation, Research and Special Programs Administration, John A. Volpe National Transportation Systems Center (Volpe Center), Acoustics Facility, in support of the Federal Highway Administration (FHWA), Office of Natural Environment, has developed the updated "FHWA Highway Noise Barrier Design Handbook." This document reflects substantial improvements and changes in noise barrier design that have evolved since the original 1976 publication. This Handbook, which is accompanied by a videotape and a companion CD-ROM, addresses both acoustical and non-acoustical issues associated with highway noise barrier design.

The objectives of this document and accompanying video and CD-ROM are to provide: (1) guidelines on how to design a highway noise barrier that fits with its surroundings and performs its intended acoustical and structural functions at reasonable life-cycle cost; and (2) a state-of-the-art reference of common concepts, designs, materials, and installation techniques for the professional highway engineer, the acoustical and design engineers and planners, and the non-professional community participant. This handbook may also be used as a guide for other applications such as noise barriers used to attenuate noise from rail lines, as well as noise from other sources which are not necessarily found in transportation. Every effort has been made to address common designs, materials, and installation techniques. However, it is impossible to encompass the proliferation of new concepts and materials entering the market on a daily basis. Therefore, the specific descriptions in this handbook are not to be considered all-inclusive, and are not intended to limit the creativeness of the designer, manufacturer, and construction contractor. Any new theory, design, material, or installation technique not addressed in this handbook should be evaluated with the general fundamentals of durability, safety, and functionality in mind.

ACKNOWLEDGEMENTS

The authors wish to express their sincere gratitude to all who helped in updating "FHWA Highway Noise Barrier Design Handbook." Special thanks go to Rudy Hendriks of CALTRANS, Win Lindeman of Florida DOT, and Ken Polcak of Maryland State Highway Administration, Domenick Billera of New Jersey DOT, and Bill McColl of New York DOT for reviewing the draft document and providing their invaluable insight and comments.

In addition, the information provided by the FHWA, State Transportation Agencies, and individuals contributed to ensuring the accuracy and level of detail of the final document. We would also like to thank the following individuals for their support and timely commentary:

Agency/Company	Location	Contact
Bowlby and Associates, Inc.	Brentwood, TN	Bill Bowlby
CALTRANS	Sacramento, CA	Rudy Hendriks
Carsonite International	Richmond, VA/Citris Heights, CA	Paul DuBay/Paul Schubring
Colorado DOT	Denver, CO	Makeba Adesunloye
Concrete Placement Systems	Chantilly, VA	Bob Glasgow
Concrete Solutions, Inc.	Austin, TX	Boone and Wendy Bucher
Connecticut DOT	Newington, CT	Carmine Trotta
Durisol	Hamilton, Canada	Hans J. Rerup
Florida DOT, District 4	Ft. Lauderdale, FL	Ken Campbell
Florida DOT, District 6	Miami, FL	Laura Letson
Florida DOT, Central Office	Tallahassee, FL	Win Lindeman
Industrial Acoustics	Bronx, NY	Gary Figallo
MD State Highway Administration	Baltimore, MD	Ken Polcak
Minnesota DOT	Saint Paul, MN	Melvin Roseen
New Jersey DOT	Trenton, NJ	Domenick Billera
New York DOT	Albany, NY	Bill McColl
North Carolina DOT	Raleigh, NC	John Alford/Steve Walker
Ohio DOT	Columbus, OH	Elvin Pinckney
Ohio University	Columbus, OH	Lloyd Herman
Oregon DOT	Salem, OR	Dave Goodwin
Pennsylvania DOT	Harrisburg, PA	Jim Byers
Powell Contracting	Richmond Hill, ON	Dwight Powell
Quebec Ministry of Transportation	Montreal, Canada	Line Gamache
Smith-Midland Company	Midland, VA	Ashley Smith
Sound Zero Corporation	Birdsboro, PA	Mark Murphy
University of Central Florida	Orlando, FL	Roger Wayson
US Gypsum	Chicago, IL	Richard Kacskowski
Virginia DOT	Richmond, VA	Cary Adkins
Wall Journal	Lehigh Acres, FL	El Angove
Wisconsin DOT	Milwaukee, WI	Jay Waldschmidt

Preceding Page Blank

TABLE OF CONTENTS

Section   **Page**

1. **INTRODUCTION** 1
   1.1 Background 1
   1.2 Objectives 1

2. **TERMINOLOGY** 3

3. **ACOUSTICAL CONSIDERATIONS** 15
   3.1 Characteristics of Sound 15
   3.2 Noise Descriptors 19
   3.3 Sound Propagation 19
       3.3.1 Divergence 19
       3.3.2 Ground Effect 20
       3.3.3 Atmospheric Effects 20
       3.3.4 Shielding by Natural and Man-Made Structures 21
   3.4 Noise Barrier Basics 22
       3.4.1 Barrier Absorption 25
       3.4.2 Barrier Sound Transmission 25
   3.5 Barrier-Design Acoustical Considerations 27
       3.5.1 Barrier Design Goals and Insertion Loss 27
       3.5.2 Barrier Length 28
       3.5.3 Wall Versus Berm 29
       3.5.4 Reflective Versus Absorptive 30
       3.5.5 Other Unique Design Considerations 32
             3.5.5.1 Overlapping Barriers 32
             3.5.5.2 "Zig-zag" Barriers 33
             3.5.5.3 Tops of Barriers 33

4. **NOISE BARRIER TYPES** 37
   4.1 Ground-Mounted 37
       4.1.1 Noise Berms 37
       4.1.2 Noise Walls 38
             4.1.2.1 Post and Panel 38
                     4.1.2.1.1 Tilted Post and Panel 43
             4.1.2.2 Brick and Masonry Block 44
             4.1.2.3 Free Standing Noise Walls 44
                     4.1.2.3.1 Precast Concrete 45
                     4.1.2.3.2 "Planted" or Bin Type Barriers 45
                     4.1.2.3.3 Stone Crib 46
             4.1.2.4 Direct Burial Panels 46
             4.1.2.5 Noise Walls Used to Partially Retain Earth 47
             4.1.2.6 Cast-In-Place Concrete Noise Walls 48
       4.1.3 Combination Noise Berm and Noise Wall Systems 48

4.2 Structure-Mounted Noise Walls ... 49
4.2.1 Noise Walls on Bridges ... 49
4.2.1.1 Types of Noise Walls on Bridges ... 50
4.2.1.2 Effect of Noise Walls on the Structural Characteristics of an Existing Bridge ... 52
4.2.1.3 Effect of Noise Wall on the Structural Requirements of a New Bridge ... 52
4.2.1.4 Potential for Damage to Noise Wall From Vehicular Impact or Airborne Debris ... 53
4.2.1.5 Potential for Damage and Injury in the Event of the Noise Wall or Parts Thereof Falling From the Structure ... 53
4.2.1.6 Other Safety-Related Concerns ... 53
4.2.1.7 Maintenance Considerations ... 54
4.2.2 Noise Walls on Retaining Walls ... 54
4.2.2.1 Combination Cast-In-Place Retaining Wall and Noise Barrier Wall ... 54
4.2.2.2 Noise Wall behind Cast-In-Place Retaining Wall ... 55
4.2.2.3 Noise Wall on or behind Retained Earth System Type Retaining Wall ... 56
4.2.2.4 Noise Barrier Walls in Combination with or behind Pre-Manufactured Retaining Wall ... 57
4.3 Special Features ... 57
4.3.1 Caps ... 57
4.3.2 Emergency Access Openings ... 57
4.3.3 Drainage Openings in Noise Walls ... 58
4.3.4 Attachments to Noise Walls ... 58

5. NOISE BARRIER MATERIALS AND SURFACE TREATMENTS ... 61
5.1 Concrete ... 61
5.1.1 Special Considerations ... 62
5.1.2 Verification of Quality ... 63
5.2 Brick and Masonry Block ... 64
5.2.1 Special Considerations ... 65
5.2.2 Verification of Quality ... 65
5.3 Metals ... 66
5.3.1 Special Considerations ... 67
5.3.2 Verification of Quality ... 68
5.4 Wood ... 68
5.4.1 Special Considerations ... 69
5.4.2 Verification of Quality ... 71
5.5 Transparent Panels ... 72
5.5.1 Special Considerations ... 72
5.5.2 Verification of Quality ... 74
5.6 Plastics ... 74
5.6.1 Special Considerations ... 75
5.6.2 Verification of Quality ... 75
5.7 Recycled Rubber ... 76
5.7.1 Special Considerations ... 76
5.7.2 Verification of Quality ... 77
5.8 Composites ... 77
5.8.1 Special Considerations ... 78
5.8.2 Verification of Quality ... 79

5.9 Barrier Surface Treatment 80
   5.9.1 Textures 80
      5.9.1.1 Concrete 81
      5.9.1.2 Masonry Block 84
      5.9.1.3 Brick 85
      5.9.1.4 Metal 86
      5.9.1.5 Wood 86
      5.9.1.6 Transparent Materials 88
      5.9.1.7 Plastics 88
      5.9.1.8 Rubber 88
      5.9.1.9 Composites 89
      5.9.1.10 Other Applications 89
      5.9.1.11 Special Considerations 89
   5.9.2 Color 90
      5.9.2.1 Concrete and Masonry Block 90
      5.9.2.2 Brick 91
      5.9.2.3 Metal 91
      5.9.2.4 Wood 92
      5.9.2.5 Plastics, Fiberglass, and Acrylics 92
      5.9.2.6 Rubber 92
      5.9.2.7 Composites 93
      5.9.2.8 Planted Walls 93
   5.9.3 Coatings 93
      5.9.3.1 Anti-Graffiti Coatings 93
      5.9.3.2 Stains 94
      5.9.3.3 Application Process 94
      5.9.3.4 Relationship of Coating Type to Maintenance Philosophy 94
      5.9.3.5 Relationship of Coating to Barrier's Acoustical Performance ... 95
      5.9.3.6 Health and Environmental Issues 95

6. NOISE BARRIER AESTHETICS 99
   6.1 Relationship of Noise Barrier to Surroundings 99
      6.1.1 Alignment Changes 99
      6.1.2 Vertical Stepping/Sloping of Panels 100
      6.1.3 Caps 102
         6.1.3.1 Horizontal Caps 102
         6.1.3.2 Vertical Caps 103
      6.1.4 Barrier End Treatments 103
      6.1.5 Special Aesthetic Considerations in Cultural/Historic Areas 104
      6.1.6 View from the Road 105
         6.1.6.1 Color 105
         6.1.6.2 Texture 106
         6.1.6.3 Pattern 106
         6.1.6.4 Shape 108
      6.1.7 View from the Adjacent Land Uses 109
         6.1.7.1 Color 109

    6.1.7.2 Texture 110
    6.1.7.3 Pattern 110
    6.1.7.4 Shape 111
 6.2 Landscaping 111
  6.2.1 Integration of Noise Barrier with Surroundings and Accommodation of Existing Vegetation 111
  6.2.2 Supplementing Existing Vegetation, Replacing Existing Vegetation, and/or Adding New Vegetation 112
  6.2.3 Consistency of Landscape Treatment with Maintenance Philosophy 114

7. DRAINAGE AND UTILITY CONSIDERATIONS 117
 7.1 Drainage Requirements 117
  7.1.1 Use of Barrier Overlap Section to Accommodate Drainage Flows 117
  7.1.2 Accommodating Water Flow Through a Barrier 117
  7.1.3 Accommodating Water Flow Along and/or Underneath a Barrier 118
  7.1.4 Special Drainage Considerations in Flood Plain Areas 119
 7.2 Lighting, Sign Supports and Utility Poles and Other Elements Mounted on or Adjacent to Noise Barriers 119
 7.3 Effects of Underground Utilities on Noise Barrier Design and Location 120
 7.4 Effects of Overhead Utilities on Noise Barrier Design and Location 120

8. STRUCTURAL CONSIDERATIONS 123
 8.1 Expansion and Contraction of Barrier Materials 123
 8.2 Noise Barrier Loadings 125
 8.3 Barrier Height Considerations 125
 8.4 Foundation Requirements 126
  8.4.1 Concrete Footings in Earth 126
  8.4.2 Concrete Footings in Rock 126

9. SAFETY CONSIDERATIONS 129
 9.1 Qualitative Evaluation of Safety 129
  9.1.1 Evaluate the Need for Special Considerations Related to Safety 129
   9.1.1.1 Probability of Occurrence of the Barrier Being Impacted 129
   9.1.1.2 Consequences of the Barrier Being Impacted 129
  9.1.2 Modifications to the Noise Barrier Design 130
  9.1.3 Overall Results of Qualitative Evaluation 131
 9.2 Sight Distance 132
 9.3 Traffic Protection 132
 9.4 Emergency Access 133
  9.4.1 Barrier Overlap Sections 133
  9.4.2 Access Doors 134
 9.5 Fire Safety 136
 9.6 Glare 138
 9.7 Shatter Resistance 139
 9.8 Icing and Snow Removal 139

10. PRODUCT EVALUATION 143
10.1 Evaluation Process 143
10.1.1 Step 1 - Submission 143
10.1.2 Step 2 - Preliminary Examination 146
10.1.3 Step 3 - Detailed Evaluation 146
10.1.4 Step 4 - Incorporation into Standards, Specifications, Manuals, and Policies 147
10.1.5 Step 5 - Implementation 147
10.1.6 Step 6 - Performance Feedback 147
10.2 Product Handling and Storage 148
10.3 Sampling and Testing Requirements 148
10.3.1 Concrete 149
10.3.2 Metals 151
10.3.3 Wood 152
10.4 Criteria for Approval 152
10.4.1 Acceptance of the Noise Barrier System Design 152
10.4.2 Acceptance of the Noise Barrier Manufacturer/Fabricator 153
10.4.3 Acceptance of Project Specific Design Details (Shop Drawings and Related Documents) 154
10.4.4 Acceptance of the Installation 155

11. INSTALLATION CONSIDERATIONS 157
11.1 Site Grading and Preparation 157
11.2 Foundation Requirements 157
11.3 Quality Assurances 157
11.3.1 Visual Examination 158
11.3.2 Proof of Certification 158
11.3.3 Testing Methods 158
11.3.4 Sampling 158
11.4 Handling and Storage of Materials on Site 159
11.5 Barrier Assembly and Construction 159
11.5.1 Construction Tolerances 159
11.5.2 Grout, Caulking, and Sealants 159
11.5.3 Anchors, Fasteners, and Lifting Inserts 159
11.5.4 Field Welds 159
11.5.5 Coatings 159
11.5.6 Installation Jigs 160
11.5.7 Installation Scaffolding 160
11.6 Construction Noise Barriers 160
11.6.1 Temporary Noise Walls and Berms 161
11.6.2 Early Construction of Permanent Noise Walls and Berms 161

12. MAINTENANCE CONSIDERATIONS 163
12.1 Repairs 163
12.2 Availability of Replacement Parts 164
12.3 Access 164
12.4 Surface/Material Wear and Deterioration 165
12.5 Landscaping 167

12.6 Graffiti	167
12.7 Litter	167
12.8 Snow Storage	168
12.9 Snow Drifting	168
12.10 Issues Related to Specific Barrier Types	168

13. COST CONSIDERATIONS ... 171
13.1 Relationship of Barrier to Project Type ... 171
    13.1.1 Noise Barrier Built as a Component of Large Construction Project ... 171
    13.1.2 Sole Noise Barrier Construction Project / Retrofit Noise Barrier Construction ... 171
13.2 Physical Conditions and Factors ... 172
    13.2.1 Accessability ... 172
    13.2.2 Transportation of Material, Equipment, and Work Force ... 172
    13.2.3 Quantity of Barrier ... 172
    13.2.4 Material Availability ... 172
    13.2.5 Weather ... 173
    13.2.6 Traffic Protection and Detours ... 173
    13.2.7 Limitation of Construction Hours ... 173
13.3 Labor Costs ... 174

14. BARRIER DESIGN PROCESS ... 177
14.1 Acoustical Evaluation ... 177
    14.1.1 Select Noise Sensitive Receivers and/or Areas for Measurement and Analysis ... 177
    14.1.2 Determine Existing Noise Levels by Measurements and/or Modeling ... 177
        14.1.2.1 Noise Measurements ... 177
        14.1.2.2 Noise Modeling ... 184
    14.1.3 Determine If There Are Any Future Noise Impacts ... 184
    14.1.4 Determine Feasibility and Reasonableness of Noise Abatement ... 185
14.2 Develop Barrier Design ... 185
    14.2.1 Community Participation ... 186

15. ASSESSING BARRIER EFFECTIVENESS ... 187
15.1 Acoustic Effectiveness ... 187
    15.1.1 Select Noise Sensitive Receivers and/or Areas for Measurement and Analysis ... 187
    15.1.2 Determine Barrier Insertion Loss by Measurements and/or Modeling ... 187
        15.1.2.1 Noise Measurements ... 189
        15.1.2.2 Noise Modeling ... 194
15.2 Non-acoustic Effectiveness ... 194
    15.2.1 Community Acceptance ... 195
    15.2.2 Cost ... 195

16. TOOLS TO ASSIST ... 197
16.1 Barrier Design Video and CD-ROM ... 197
16.2 FHWA Traffic Noise Model ... 197
16.3 AASHTO, ANSI, ASTM, CSA, IEC, ISO, NCHRP, NIST, and SAE Standards ... 198
16.4 Technical Documents ... 200

16.5 State Departments of Transportation 201
16.6 Training Courses 205

REFERENCES 207

INDEX 215

LIST OF FIGURES

Figure   **Page**

Cover 1. (data base #3042)
Cover 2. (data base #2986)
Cover 3. (data base #1583)
Cover 4. (data base #8051)
Cover 5. (data base #3126)
Figure 1. Example of continuous footing. 4
Figure 2. Example of slope hinge point. 11
Figure 3. Graphical representation of L_{AE}. 12
Figure 4. Example of spread footing. 13
Figure 5. Decibel scale. 15
Figure 6. Sound wave amplitude and wavelength. 17
Figure 7. Frequency A-weighting. 18
Figure 8. Barrier absorption, transmission, reflection, and diffraction. 22
Figure 9. Barrier diffraction. 23
Figure 10. Path length difference. 23
Figure 11. Barrier attenuation versus Fresnel Number. 24
Figure 12. Barrier absorption: Reverberation Room Method (data base #2553). 25
Figure 13. Line-of-sight. 28
Figure 14. Barrier length. 29
Figure 15. Barrier curved inward toward the community (data base #2617). 29
Figure 16. Wall, berm, and combination noise barriers. 30
Figure 17. Reflective noise paths due to a single barrier. 30
Figure 18. Parallel noise barriers (data base #2968). 31
Figure 19. Reflective noise paths due to a parallel barrier. 31
Figure 20. Overlapping barriers (data base #5902). 32
Figure 21. Overlapping barriers. 33
Figure 22. "Zig-zag" barriers (data base #8057). 33
Figure 23. Special acoustical considerations: tops of barriers. 33
Figure 24. Special top of barrier (database #2395). 34
Figure 25. T-profile top barrier (database #1312). 34
Figure 26. Noise berm (data base #2584). 37
Figure 27. Noise berm (data base #2712). 37
Figure 28. Post and panel noise wall (data base #27). 38
Figure 29. Post and panel attachments: concrete cylinder (data base #528). 39
Figure 30. Post and panel attachments: continuous footing (data base #529). 39
Figure 31. Post and panel attachments: embedded post (data base #5). 39
Figure 32. Full height panel (data base #2946). 40
Figure 33. Stacked panel (data base #2680). 40
Figure 34. Post and panel: shipping requirements (data base #527). 40
Figure 35. Post and panel: panel aesthetics (data base #2979). 41

Figure 36. Post and panel: installation implications (data base #6049). 41
Figure 37. Post and panel: maintenance considerations (data base #5572). 42
Figure 38. Tilted post and panel noise wall: community side (data base #36). 43
Figure 39. Tilted post and panel noise wall: highway side (data base #34). 43
Figure 40. Brick noise wall (data base #8014). 44
Figure 41. Brick noise wall (data base #560). 44
Figure 42. Masonry block noise wall (data base #2454). 44
Figure 43. Masonry block noise wall (data base #2457). 44
Figure 44. Precast free standing concrete noise wall (data base #1206). 45
Figure 45. Precast free standing concrete noise wall (data base #1218). 45
Figure 46. Bin type noise wall: plastic (data base #247). 46
Figure 47. "Planted" noise wall: concrete (data base #2567). 46
Figure 48. Stone crib noise wall (data base #549). 46
Figure 49. Direct burial panels (data base #348). 47
Figure 50. Direct burial panels (data base #351). 47
Figure 51. Noise walls used to partially retain earth (data base #8027). 47
Figure 52. Noise walls used to partially retain earth (data base #480). 47
Figure 53. Cast-in-place concrete noise wall (data base #1054). 48
Figure 54. Combination noise berm and noise wall system (data base #27). 49
Figure 55. Combination noise berm and noise wall system (data base #993). 49
Figure 56. Noise wall on a bridge (data base #2374). 49
Figure 57. Noise wall on a bridge (data base #5231). 50
Figure 58. Noise wall on a bridge (data base #1717). 50
Figure 59. Noise wall on a bridge (data base #5090). 50
Figure 60. Noise wall damage from vehicular impact (data base #5331). 53
Figure 61. Noise wall damage from airborne debris (data base #1281). 53
Figure 62. Noise wall on a retaining wall (data base #2947). 54
Figure 63. Noise wall on a retaining wall (data base #531). 54
Figure 64. Combination cast-in-place retaining wall and noise wall (data base #1844). 54
Figure 65. Combination cast-in-place retaining wall and noise wall (data base #5004). 55
Figure 66. Combination cast-in-place retaining wall and noise wall (data base #329). 55
Figure 67. Noise wall behind cast-in-place retaining wall (data base #1745). 56
Figure 68. Noise wall behind cast-in-place retaining wall (data base #1691). 56
Figure 69. Noise wall caisson foundations behind pre-manufactured retaining wall (data base #6534). ... 57
Figure 70. Concrete noise wall (data base #634). 61
Figure 71. Concrete noise wall (data base #1239). 61
Figure 72. Brick noise barrier (data base #6511). 65
Figure 73. Masonry block noise barrier (data base #2456). 65
Figure 74. Barrier concrete foundation (data base #568). 65
Figure 75. Scaffolding for barrier installation (data base #2455). 65
Figure 76. Metal noise barrier (data base #158). 66
Figure 77. Metal noise barrier (data base #5692). 66
Figure 78. Metal barrier: climbability (data base #157). 67
Figure 79. Plywood noise barrier (data base #657). 69
Figure 80. Glue laminated post and plank noise barrier (data base #736). 69
Figure 81. Wood barrier: tongue and groove planking (data base #411). 70

Figure 82. Transparent panel noise barrier (data base #1981). 72
Figure 83. Transparent panel barrier: vandalism (data base #1947). 72
Figure 84. Plastic noise barrier (data base #782). 74
Figure 85. Recycled rubber noise barrier (data base #3124). 76
Figure 86. Composite noise barrier (data base #132). 78
Figure 87. Composite noise barrier (data base #707). 78
Figure 88. Barrier surface treatment: textures (data base #512). 80
Figure 89. Concrete: smooth surface (data base #996). 81
Figure 90. Concrete: exposed aggregate (data base #748). 81
Figure 91. Concrete: form liner (data base #1180). 82
Figure 92. Concrete: form liner (data base #498). 82
Figure 93. Concrete: form liner (data base #698). 82
Figure 94. Concrete: form liner panel joints (data base #7029). 82
Figure 95. Concrete: raked finish (data base #508). 83
Figure 96. Concrete: stamped finish (data base #6512). 83
Figure 97. Concrete: stamped finish (data base #6514). 83
Figure 98. Concrete: inserts (data base #6532). 84
Figure 99. Concrete: inserts (data base #8015). 84
Figure 100. Concrete: stucco (data base #1066). 84
Figure 101. Masonry block: fractured fin (data base #948). 85
Figure 102. Masonry block: stucco (data base #1061). 85
Figure 103. Brick: surface texture (data base #560). 85
Figure 104. Brick: surface texture (data base #6518). 85
Figure 105. Metal: surface texture (data base #1708). 86
Figure 106. Wood: horizontal plank orientation (data base #745). 86
Figure 107. Wood: vertical plank orientation (data base #730). 86
Figure 108. Wood: patterns on battens attached to panels (data base #471). 87
Figure 109. Wood: pattern on laminated panels (data base #663). 87
Figure 110. Wood: circular post type (data base #744). 87
Figure 111. Wood: square post type (data base #435). 87
Figure 112. Transparent: surface stencil design (data base #1954). 88
Figure 113. Plastic: surface texture (data base #792). 88
Figure 114. Rubber: surface texture (data base #2948). 88
Figure 115. Rubber: surface texture (data base #2949). 88
Figure 116. Composite: surface texture (data base #136). 89
Figure 117. Composite: surface texture (data base #708). 89
Figure 118. Gunite: surface texture (data base #2243). 89
Figure 119. Surface texture: special considerations - stacked panel joint (data base #902). 90
Figure 120. Surface texture: special considerations - form liner joint (data base #533). 90
Figure 121. Concrete: color (data base #2337). 90
Figure 122. Concrete: color (data base #1218). 91
Figure 123. Masonry block: color (data base #2373). 91
Figure 124. Brick: color (data base #560). 91
Figure 125. Metal: color (data base #1720). 91
Figure 126. Wood: color (data base #464). 92
Figure 127. Plastic: color (data base #1728). 92

Figure 128. Composites: color (data base #2533). 93
Figure 129. Anti-graffiti coating (data base #5332). 94
Figure 130. Alignment changes (data base #1496). 100
Figure 131. Alignment changes (data base #6524). 100
Figure 132. Alignment changes: possible flanking reflections (data base #8030). 100
Figure 133. Vertical stepping of panels: uniform (data base #6523). 100
Figure 134. Vertical sloping of panels. 101
Figure 135. Vertical stepping of panels: irregular (data base #111). 101
Figure 136. Vertical sloping of panels: smooth (data base #1839). 101
Figure 137. Noise wall horizontal cap (data base #271). 102
Figure 138. Noise wall horizontal cap (data base #1325). 102
Figure 139. Noise wall horizontal cap (data base #2434). 102
Figure 140. Noise wall horizontal cap (data base #1731). 102
Figure 141. Noise wall vertical cap (data base #2541). 103
Figure 142. Noise wall vertical cap: damage (data base #5224). 103
Figure 143. Barrier end treatment: buried into existing ground (data base #80). 103
Figure 144. Barrier end treatment: stepped panel (data base #193). 104
Figure 145. Barrier end treatment: sloped panel (data base #2385). 104
Figure 146. Barrier end treatment: vegetation (data base #1243). 104
Figure 147. Barrier end treatment: berming (data base #1270). 104
Figure 148. Special considerations in cultural areas (data base #6533). 104
Figure 149. Special considerations in historic areas (data base #6516). 104
Figure 150. View from the road (data base #2007). 105
Figure 151. View from the road (data base #3128). 105
Figure 152. View from the road: color (data base #3118). 105
Figure 153. View from the road: color (data base #1218). 105
Figure 154. View from the road: texture (data base #2368). 106
Figure 155. View from the road: texture (data base #652). 106
Figure 156. View from the road: pattern (data base #1122). 106
Figure 157. View from the road: pattern (data base #453). 107
Figure 158. View from the road: pattern (data base #2296). 107
Figure 159. View from the road: pattern (data base #2316). 107
Figure 160. View from the road: pattern (data base #2352). 107
Figure 161. View from the road: pattern (data base #328). 107
Figure 162. View from the road: pattern (data base #5386). 107
Figure 163. View from the road: shape (data base #1238). 108
Figure 164. View from the road: shape (data base #1190). 108
Figure 165. View from adjacent land uses (data base #148). 109
Figure 166. View from adjacent land uses: color (data base #7039). 109
Figure 167. View from adjacent land uses: color (data base #8069). 109
Figure 168. View from adjacent land uses: texture (data base #653). 110
Figure 169. View from adjacent properties: pattern (data base #458). 110
Figure 170. View from adjacent properties: pattern (data base #2420). 110
Figure 171. View from adjacent properties: pattern (data base #2562). 110
Figure 172. View from adjacent properties: pattern (data base #2309). 110
Figure 173. View from adjacent properties: shape (data base #664). 111

Figure 174. Landscaping: integration with existing vegetation (data base #2560). 112
Figure 175. Landscaping: integration with existing vegetation (data base #6526). 112
Figure 176. Landscaping: supplementing vegetation (data base #1975). 112
Figure 177. Landscaping: supplementing vegetation (data base #6530). 112
Figure 178. Landscaping: supplementing vegetation (data base #470). 113
Figure 179. Landscaping: supplementing vegetation (data base #820). 113
Figure 180. Landscaping: supplementing vegetation (data base #51). 113
Figure 181. Landscaping: supplementing vegetation (data base #1759). 113
Figure 182. Landscaping: blocking panel aesthetic features (data base #1212). 114
Figure 183. Landscaping: consistency with maintenance philosophy (data base #6531). 115
Figure 184. Landscaping: consistency with maintenance philosophy (data base #2155). 115
Figure 185. Drainage: use of barrier overlap (data base #901). 117
Figure 186. Drainage: use of barrier overlap (data base #382). 117
Figure 187. Drainage: water through a barrier (data base #1414). 118
Figure 188. Drainage: water through a barrier (data base #1132). 118
Figure 189. Drainage: water along a barrier (data base #984). 118
Figure 190. Drainage: water along a barrier (data base #1065). 118
Figure 191. Drainage: water underneath a barrier (data base #1798). 118
Figure 192. Drainage: water underneath a barrier (data base #1799). 118
Figure 193. Drainage: water underneath a barrier (data base #2071). 119
Figure 194. Drainage: water through a swing panel (data base #6168). 119
Figure 195. Lighting incorporated into a barrier system (data base #1180). 119
Figure 196. Other elements mounted on a barrier (data base #1074). 120
Figure 197. Other elements mounted on a barrier (data base #1432). 120
Figure 198. Effects of underground utilities on barrier design (data base #1089). 120
Figure 199. Effects of underground utilities on barrier design (data base #2228). 120
Figure 200. Effects of overhead utilities on barrier design (data base #835). 121
Figure 201. Effects of overhead utilities on barrier design (data base #5288). 121
Figure 202. Effects of overhead utilities on barrier design (data base #869). 121
Figure 203. Expansion and contraction of materials: post to panel connections (data base #8038). 123
Figure 204. Expansion and contraction of materials: panel to panel connections (data base #181). 123
Figure 205. Expansion and contraction of barrier materials: connections between barriers
(data base #463). 124
Figure 206. Expansion and contraction of barrier materials: connections between barriers
(data base #1716). 124
Figure 207. Expansion and contraction of barrier materials: connections between barriers
(data base #1720). 124
Figure 208. Expansion and contractions of barrier materials: structure barriers (data base #414). 124
Figure 209. Expansion and contraction of barrier materials: structure barriers (data base #1711). 124
Figure 210. Stepped concrete footings in earth (data base #2939). 126
Figure 211. Consequences of a barrier being impacted (data base #1148). 130
Figure 212. Barrier attachment/reinforcement details (data base #1267). 130
Figure 213. Sight distance (data base #612). 132
Figure 214. Traffic protection (data base #1251). 132
Figure 215. Traffic protection (data base #64). 132
Figure 216. Traffic protection (data base #4). 133

Figure 217. Traffic protection (data base #8036).	133
Figure 218. Barrier overlap sections (data base #109).	134
Figure 219. Barrier overlap sections (data base #382).	134
Figure 220. Access door (data base #1519).	134
Figure 221. Access door (data base #1631).	134
Figure 222. Access door (data base #6519).	134
Figure 223. Access door (data base #1539).	135
Figure 224. Access door (data base #2344).	135
Figure 225. Access door (data base #2414).	135
Figure 226. Emergency access opening (data base #5017).	135
Figure 227. Emergency access opening (data base #2290).	135
Figure 228. Access for fire (data base #1416).	136
Figure 229. Access for fire (data base #54).	136
Figure 230. Access for fire (data base #260).	136
Figure 231. Access for fire (data base #8089).	137
Figure 232. Access for fire (data base #1714).	137
Figure 233. Access for fire (data base #1568).	137
Figure 234. Access for fire (data base #1112).	137
Figure 235. Glare (data base #392).	138
Figure 236. Installation jigs (data base #6536).	160
Figure 237. Installation considerations (data base #6535).	160
Figure 238. Installation considerations (data base #2406).	160
Figure 239. Installation considerations (data base #2327).	160
Figure 240. Construction noise barriers (data base #6539b).	161
Figure 241. Early construction of permanent noise barriers (data base #6538).	161
Figure 242. Repairs (data base #736a).	163
Figure 243. Repairs (data base #2488).	163
Figure 244. Repairs (data base #1052).	164
Figure 245. Repairs (data base #450).	164
Figure 246. Availability of replacement parts (data base #226).	164
Figure 247. Deterioration from moisture (data base #449).	165
Figure 248. Deterioration from moisture (data base #1008).	165
Figure 249. Deterioration from ultraviolet light (data base #1501).	166
Figure 250. Graffiti (data base #757).	167
Figure 251. Litter (data base #1642).	167
Figure 252. Snow storage (data base #5342).	168
Figure 253. Snow drifting (data base #2540).	168
Figure 254. Issues related to specific barrier types (data base #3125).	168
Figure 255. Generic measurement instrumentation setup.	178
Figure 256. Reference microphone - position 1.	190
Figure 257. Reference microphone - position 2.	190
Figure 258. Receiver positions.	191

LIST OF TABLES

Table   **Page**

Table 1. Decibel addition approximation. 16
Table 2. Frequency A-weighting. 18
Table 3. Approximate sound transmission loss values for common materials 26
Table 4. Relationship between barrier insertion loss and design feasibility 28
Table 5. Guideline for categorizing parallel barrier sites based on w/h ratio 32
Table 6. Sampling period. 182
Table 7. Classes of wind conditions. 188
Table 8. Classes of cloud cover. 189

1. INTRODUCTION

The U.S. Department of Transportation, Research and Special Programs Administration, John A. Volpe National Transportation Systems Center (Volpe Center), Acoustics Facility, in support of the Federal Highway Administration (FHWA), Office of Natural Environment, has developed the updated "FHWA Highway Noise Barrier Design Handbook." This document reflects substantial improvements and changes in noise barrier design that have evolved since the original 1976 publication.[1] This Handbook, which is accompanied by a videotape and a companion CD-ROM, addresses both acoustical and non-acoustical issues associated with highway noise barrier design. [2,3]

Section 1 presents a general overview, a historical perspective, and the objectives of the handbook. Section 2 presents definitions of terminology used throughout the document. Section 3 describes the acoustical considerations of highway noise barrier design, including a brief discussion on noise barrier performance. Section 4 presents noise barrier types, their descriptions, and special features. Section 5 describes noise barrier materials, including barrier surface texture treatments. Section 6 discusses noise barrier aesthetics. Section 7 describes the drainage and utility considerations associated with barrier design. Section 8 describes the structural considerations associated with barrier design. Section 9 describes the safety considerations associated with barrier design. Section 10 details the product evaluation process. Section 11 describes the installation considerations associated with barrier design. Section 12 describes the maintenance considerations associated with barrier design. Section 13 describes the cost considerations associated with barrier design. Section 14 presents the typical barrier design process. Section 15 describes how to assess a barrier's effectiveness, including performance, costs, and community acceptance. Section 16 describes the various tools and information sources that are available to aid in the design process.

1.1 Background

Highway traffic noise has been a Federal, State, and local concern since the first noise barrier was built in 1963. In 1976, the FHWA developed the original "Noise Barrier Design Handbook" to aid State Highway Agencies in solving the problem of highway traffic noise. However, the Handbook was written over two decades ago. Since then, substantial advancements in the methodology and technology of barrier design have occurred in concept, design, and technique. Increased community and motorist interest has fueled the push to provide better, less expensive, and more environmentally friendly barrier designs.

Increased community and motorist concerns have also fueled the push to improve noise measurement and modeling technologies which aid State transportation agencies in determining which communities need noise abatement. One such tool is the FHWA's recently released highway traffic noise prediction model: the FHWA Traffic Noise Model (FWHA TNM®).[4,5,6,7] The FHWA TNM is an entirely-new, state-of-the-art computer program used for predicting noise impacts in the vicinity of highways. It uses advances in personal computer hardware and software to improve upon the accuracy and ease of modeling highway traffic noise, including the design of effective, cost-efficient highway noise barriers.

1.2 Objectives

The objectives of this document and accompanying video and CD-ROM are to provide: (1) guidelines on how to design a highway noise barrier that fits with its surroundings and performs its intended acoustical and structural functions at reasonable life-cycle cost; and (2) a state-of-the-art reference of common concepts, designs, materials, and installation techniques for the professional highway engineer, the noise barrier designer, and the non-professional community participant. This handbook may also be used as a guide for other applications such as noise barriers used to attenuate noise from rail lines, as well as noise from other sources which are not necessarily found in transportation. Every effort has been made to address common designs, materials, and installation techniques. However, it is impossible to encompass the proliferation of new concepts and materials entering the market on a daily basis. Therefore, the specific descriptions in this handbook are not to be considered all-inclusive, and are not intended to limit the creativeness of the designer, manufacturer, and construction contractor. Any new theory, design, material, or installation technique not addressed in this handbook should be evaluated with the general fundamentals of durability, safety, and functionality in mind.

2. TERMINOLOGY

This section presents pertinent terminology used throughout the document. These terms are highlighted with boldface type when they first appear in subsequent sections. Note: Definitions are generally consistent with those of References 8 through 12.

A-WEIGHTING: The weighting network used to account for changes in level sensitivity as a function of frequency. The A-weighting network de-emphasizes the high (6.3 kHz and above) and low (below 1 kHz) frequencies, and emphasizes the frequencies between 1 kHz and 6.3 kHz, in an effort to simulate the relative response of the human ear. See also frequency weighting.

ACOUSTIC ENERGY: Commonly referred to as the mean-square sound-pressure ratio, sound energy, or just plain energy, acoustic energy is the square of the ratio of the mean-square sound pressure (often frequency weighted), and the reference mean-square sound pressure of 20 µPa, the threshold of human hearing. It is arithmetically equivalent to $10^{(SPL/10)}$, where SPL is the sound pressure level, expressed in decibels.

AMBIENT NOISE: All-encompassing sound that is associated with a given environment, usually a composite of sounds from many sources near and far.

AMPLITUDE: The maximum value of a sinusoidal quantity measured from peak to peak.

ANCHOR: A bolt, stud, or reinforcing bar embedded in concrete.

ARTIFICIAL NOISE SOURCE: An acoustical source that is controlled in position and calibrated as to output power, spectral content, and directivity.

BACKER ROD: Flexible foam polyethylene rope manufactured in various diameters and suitable for use as seals and joint fillers.

BACKFILL: The envelope of engineered soil, excluding the bedding, placed around the footing or foundation of a structure in a controlled manner.

BACKGROUND NOISE: All-encompassing sound of a given environment without the sound source of interest.

BAY: The area between two posts in a noise barrier system. This area can contain full height or stacked panels, cast-in-place panels, or other types of assembled panel components.

BEDDING: The prepared portion of the engineered soil on which the footing or foundation of a structure is placed.

BRIDGE: A structure which provides a roadway, or walkway, for the passage of vehicles and pedestrians across an obstruction, gap, water course, or facility and which is typically greater than 3 m (10 ft) in span.

CAISSON: A wood, metal, or concrete casing sunk or constructed below ground or water level.

CHECK (WOOD): A separation of the wood along the grain, the greater part of which occurs across the rings of annual growth.

COATING HOLIDAY: Unwanted breaks or interruptions in coating integrity.

COMMUNITY NOISE EQUIVALENT LEVEL (CNEL, denoted by the symbol, L_{den}): A 24-hour time-averaged L_{AE} (see definition on Page 12), adjusted for average-day sound source operations. In the case of highway traffic noise, a single operation is equivalent to a single vehicle pass-by. The adjustment includes a 5-dB penalty for vehicle pass-bys occurring between 1900 and 2200 hours, local time, and a 10-dB penalty for those occurring between 2200 and 0700 hours, local time. The L_{den} noise descriptor is used primarily in the State of California. L_{den} is computed as follows:[*]

$$L_{den} = L_{AE} + 10*\log_{10}(N_{day} + 3*N_{eve} + 10*N_{night}) - 49.4 \qquad (dB)$$

where:
L_{AE} = Sound exposure level in dB (see definition on Page 12);
N_{day} = Number of vehicle pass-bys between 0700 and 1900 hours, local time;
N_{eve} = Number of vehicle pass-bys between 1900 and 2200 hours, local time;
N_{night} = Number of vehicle pass-bys between 2200 and 0700 hours, local time; and
49.4 = A normalization constant which spreads the acoustic energy associated with highway vehicle pass-bys over a 24-hour period, i.e., $10*\log_{10}(86,400$ seconds per day$) = 49.4$ dB.

COMPACTION: The process of soil densification, at a specified moisture content, due to application of loads through rolling, kneading, tamping, rodding, or vibratory actions of mechanical or manual equipment.

CONCRETE COVER: The least distance between the surface of the reinforcing bar, mesh, wire, strands or post tensioning ducts, and the surface of the concrete.

CONTINUOUS FOOTINGS: A reinforced concrete beam running the entire length of the noise barrier panels set on, or just below, the ground line (see Figure 1). These footings are designed to support and evenly distribute the dead load of the entire length of each noise barrier section or panel.

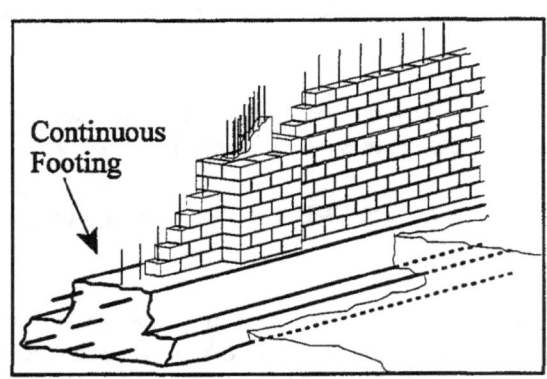

Figure 1. Example of continuous footing.

---

[*] In accordance with the technical definition, a 5-dB penalty is added to evening operations when computing the L_{den} noise metric. The 5-dB penalty, expressed in terms of a weighting factor, is equivalent to 3.16 not 3. However, in Title 21 Subchapter 6, 5001 of California state law, a factor of 3 is used. Since the State of California is the primary user of the L_{den} metric, the computation of the metric is consistent with state law, rather than the traditional technical definition.

CONTOUR: Graphical plot consisting of a smooth curve, statistically regressed through points of equal level.

CRACK (WOOD): A separation of the wood cells across the grain (this may be due to internal strains resulting from unequal longitudinal shrinkage or to external forces).

CREEP: Time-dependent deformation of a material under a sustained load.

CULVERT: A structure which provides an opening through an embankment for the purpose of the passage of water and in which roadway or structural loads are distributed to the culvert structure through fill.

DAY-NIGHT AVERAGE SOUND LEVEL (DNL, denoted by the symbol, L_{dn}): A 24-hour time-averaged L_{AE} (see definition on Page 12), adjusted for average-day sound source operations. In the case of highway traffic noise, a single operation is equivalent to a single vehicle pass-by. The adjustment includes a 10-dB penalty for vehicle pass-bys occurring between 2200 and 0700 hours, local time. L_{dn} is computed as follows:

$$L_{dn} = L_{AE} + 10*\log_{10}(N_{day} + 10*N_{night}) - 49.4 \quad \text{(dB)}$$

where:
L_{AE} = Sound exposure level in dB (see definition on Page 12);
N_{day} = Number of vehicle pass-bys between 0700 and 1900 hours, local time;
N_{night} = Number of vehicle pass-bys between 1900 and 0700 hours, local time; and
49.4 = A normalization constant which spreads the acoustic energy associated with highway vehicle pass-bys over a 24-hour period, i.e., $10*\log_{10}(86,400$ seconds per day$) = 49.4$ dB.

DEAD LOAD: The weight of all material supported by the structure and not subject to movement.

DECAY (ROT): The disintegration of the wood substance, due to the action of wood-destroying fungi.

DECIBEL (dB): A unit of measure of sound level. The number of decibels is calculated as ten times the base-10 logarithm of the square of the ratio of the mean-square sound pressure (often frequency weighted), and the reference mean-square sound pressure of 20 μPa, the threshold of human hearing.

DEGRADATION: The increase in noise levels at receivers due to conditions such as reflections from a single barrier, multiple reflections of the noise between parallel barriers, noise leaks in a barrier, etc.

DELAMINATION: A fracture plane below the surface of concrete or other material, typically parallel with the surface.

DIFFRACTED WAVE: A sound wave whose front has been changed in direction by an obstacle in the propagation medium, where the medium is air for the purposes of this document.

DIRECT EMBEDMENT: A support (the lower end of which is placed in an excavated hole and then backfilled with soil) made of concrete or other material.

DISTRESS: Excessive cracking or deformation.

DIVERGENCE: The spreading of sound waves from a source in a free field environment. In the case of highway traffic noise, two types of divergence are common, spherical and cylindrical. Spherical divergence is that which would occur for sound emanating from a point source; e.g., a single vehicle pass-by. Cylindrical divergence is that which would occur for sound emanating from a line source, or many point sources sufficiently close to behave as a line source; e.g., a continuous stream of roadway traffic.

ENERGY: See Acoustic energy.

ENGINEERED SOIL: Placed soil of known geotechnical properties.

EQUIVALENT SOUND LEVEL (TEQ, denoted by the symbol, L_{AeqT}): Ten times the base-10 logarithm of the square of the ratio of time-mean-square, instantaneous A-weighted sound pressure, during a stated time interval, T (where $T=t_2-t_1$), divided by the squared reference sound pressure of 20 µPa, the threshold of human hearing; e.g., 1HEQ, denoted by the symbol, L_{Aeq1H}, represents the hourly equivalent sound level. L_{AeqT} is related to L_{AE} by the following equation:

$$L_{AeqT} = L_{AE} - 10*\log_{10}(t_2-t_1) \qquad (dB)$$

where L_{AE} = Sound exposure level in dB (see definition on Page 12).

EXISTING LEVEL: The measured or calculated existing noise level at a given location.

EXPANSION COEFFICIENT: The rate at which specific materials expand and contract as a result of changes in temperature.

EXPONENTIAL TIME-AVERAGING: A method of stabilizing instrumentation response to signals with changing amplitude over time using a low-pass filter with a known, electrical time constant. The time constant is defined as the time required for the output level to reach 63.4 percent of the input, assuming a step-function input. Also, the output level will typically reach 100 percent of an input-step function after approximately five time constants.

FAR FIELD: That portion of a point source's sound field in which the sound pressure level (due to this sound source) decreases by 6 dB per doubling of distance from the source; i.e., spherical divergence. For a line source, the far-field is the portion of the sound field in which the sound pressure level decreases by 3 dB per doubling of distance.

FLASHING: A metal, plastic, or fabric sheeting used to cover unavoidable gaps between noise barrier components and to prevent water from entering cavities, gaps, joints, or cracks in a noise barrier wall system, which may cause premature deterioration of the noise barrier components.

FLUTING: A texture produced on panels by incorporating ribbed form liners into the molds.

FOUNDATION: That part of a structure, or bridge substructure, that transfers loads to the soil, rock, or engineered fill.

FREE FIELD: A sound field whose boundaries exert a negligible influence on the sound waves. In a free-field environment, sound spreads spherically from a source and decreases in level at a rate of 6 dB per doubling of distance from a point source, and at a rate of 3 dB per doubling of distance from a line source.

FREQUENCY: The number of cycles of repetition per second or the number of wavelengths that have passed by a stationary point in one second.

FREQUENCY WEIGHTING: A method used to account for changes in sensitivity as a function of frequency. Three standard weighting networks, A, B, and C, are used to account for different responses to sound pressure levels. Note: The absence of frequency weighting is referred to as "flat" response. See also A-weighting.

FRESNEL NUMBER: A dimensionless value used in predicting the attenuation value used in predicting the attenuation provided by a noise barrier positioned between a source and a receiver.

FROST HEAVE: A seasonal up-thrust of the ground, or pavement, caused by the formation of ice layers in a frost susceptible soil.

GABION: (or Stone Crib) A wire mesh basket filled with stones, or broken concrete, which forms part of a larger unit used for slope stability, erosion control, or for other purposes, such as a wall noise barrier.

GIRTS: The metal beams attached between the posts forming a framework to support the anchoring of metal panels to the structure.

GLUE-LAMINATED WOOD/TIMBER: A structural wood component produced by gluing together a number of laminations having their grain essentially parallel.

GRADE: The slope of the roadway, or roadway segment (expressed in percent). For example, a roadway that is 400 m in length and its end is 20 m higher in elevation relative to its start, has a 5-percent grade; i.e., 20/400 * 100 percent.

GROUND EFFECT: The change in sound level, either positive or negative, due to intervening ground between source and receiver. Ground effect is a relatively complex acoustic phenomenon, which is a function of ground characteristics, source-to-receiver geometry, and the spectral characteristics of the source. A commonly used rule-of-thumb for propagation over soft ground (e.g., grass) is that ground effects will account for about 1.5 dB per doubling of distance. However, this relationship is quite empirical and tends to break down for distances greater than about 30.5 to 61 m (100 to 200 ft).

GROUND IMPEDANCE: A complex function of frequency relating the sound transmission characteristics of a ground surface type. Measurements to determine ground impedance must be made in accordance with the ANSI standard for measuring ground impedance, scheduled for publication in 1999.[13]

HARD GROUND: Any highly reflective surface in which the phase of the sound energy is essentially preserved upon reflection; examples include water, asphalt, and concrete.

ICE CREDATION: The build-up of an ice layer on the exposed surfaces of a body due to freezing rain or in-cloud icing.

IMPACT CRITERION LEVEL: The level defined by a State Highway Agency (SHA). It should be at least 1 dB(A) less than FHWA's appropriate Noise Abatement Criterion.

INCISING: The process of inserting and withdrawing a series of closely-spaced knife points into the surface of wood prior to pressure preservative treatment. This promotes deeper penetrations of the treatment.

INSECT DAMAGE: The damage resulting from boring insects or insect larvae.

INSERTION LOSS (IL): The sound level at a given receiver before the construction of a barrier minus the sound level at the same receiver after the construction of the barrier. The construction of a noise barrier usually results in a partial loss of soft-ground attenuation. This is due to the barrier forcing the sound to take a higher path relative to the ground plane. Therefore, barrier IL is the net effect of barrier diffraction, combined with this partial loss of soft-ground attenuation.

KERF: Small incisions cut into the surface of wood products to allow deeper penetration of the preservative solutions during the pressure-treating process.

KNOT: That portion of a branch or limb that has become incorporated in the body of a tree.

L_{AE}: See Sound exposure level.

L_{AeqT}: See Equivalent sound level.

L_{den}: See Community noise equivalent level.

L_{dn}: See Day-night average sound level.

L_{10}: See Ten-percentile exceeded sound level.

LINE-OF-SIGHT: Refers to the direct path from the source to receiver without any intervening objects or topography.

LINE SOURCE: Multiple point sources moving in one direction; e.g., a continuous stream of roadway traffic, radiating sound cylindrically. Note: Sound levels measured from a line source decrease at a rate of 3 dB per doubling of distance.

LIVE LOAD: A load imposed by vehicles, pedestrians, equipment, or components subject of movement, other than a collision load.

LOAD FACTOR: A factor applied to loads to take into account the variability of loads, the lack of precision in the analysis of load effects, and the reduced probability of loads from different sources acting simultaneously.

LOWER BOUND TO INSERTION LOSS: The value reported for insertion loss when background levels are not measured or are too high to determine the full attenuation potential of the barrier.

MANDRELS: The center pin used in sheet metal rivets to expand and anchor the rivet body in place.

MAXIMUM SOUND LEVEL (MXFA or MXSA, denoted by the symbol, L_{AFmx} or L_{ASmx}): The maximum, A-weighted sound level associated with a given event (see Figure 2 on Page 12). Fast-scale response (L_{AFmx}) and slow-scale response (L_{ASmx}) characteristics effectively damp a signal as if it were to pass through a low-pass filter with a time constant of 125 and 1000 milliseconds, respectively. Note: Fast response is typically used for measuring individual highway vehicle pass-bys. Slow response is recommended for the measurement of long-term impact due to highway traffic noise, where impulsive noises are not dominant, and is also used for measurements of sound source levels which vary slowly as a function of time, such as aircraft.

NEAR FIELD: The sound field between the source and the far field. The near field exists under optimal conditions at distances less than four times the largest sound source dimension.

NOISE: Any unwanted sound. "Noise" and "sound" are used interchangeably in this document.

NOISE BARRIER: The structure, or structure together with other material, that potentially alters the noise at a site from a BEFORE condition to an AFTER condition.

NOISE REDUCTION COEFFICIENT (NRC): A single-number rating of the sound absorption properties of a material; it is the arithmetic mean of the Sabine absorption coefficients (see below) at 250, 500, 1000, and 2000 Hz, rounded to the nearest multiple of 0.05.

NOISE REDUCTION GOAL: The amount of noise reduction that is desired. This value should be defined by a respective State Highway Agency and should typically be in the range of 5-10 dB(A). Noise barriers must provide at least a 5-dB(A) reduction in highway traffic noise levels in order to provide noticeable and effective attenuation. A noise barrier should be designed to achieve the greatest reduction possible, but in no instance less than 5 dB(A).

NONDESTRUCTIVE TESTING: The determination of the physical properties of a component without impairment of, or removal of any material, excluding the necessary surface preparation for testing.

NORMAL INCIDENT (SOUND): (Also referred to as 0-degrees incidence.) Sound waves that strike a receiver at an angle perpendicular, or normal, to the angle of incidence.

OIL CANNING: A moderate deformation or buckling of sheet material, particularly common with flat sheet metal surfaces and typically caused by uneven stresses at the fastening points. This terminology also refers to the popping sound made when pressure is applied to the deformed sheet forcing the deformation in the opposite direction.

PANEL: The component of a barrier, which when joined together, produces a solid wall.

PARALLEL BARRIER: The condition where two noise barriers flank a roadway; i.e., one on each side.

PARAPET: Walls, railings, or a combination of both located along the outside edges of a bridge deck and designed to prevent vehicles from running off the sides of the bridge.

PERTURBATION: The height increment that a noise barrier's input height is increased (perturbed up) or decreased (perturbed down) during the barrier design process.

PILE: A relatively slender deep foundation unit, wholly or partly embedded in the ground, installed by driving, drilling, auguring, jetting, or otherwise, and which derives its capacity from the surrounding soil and/or from the soil or rock strata below its tip.

POINT LOADS: Loads generated when loads are unevenly distributed and concentrated on a very small area.

POINT SOURCE: Source that radiates sound spherically. Note: Sound levels measured from a point source decrease at a rate of 6 dB per doubling of distance.

POSTS: The component of a noise barrier, which vertically supports the barrier panels.

PRECAST MEMBERS: Concrete elements cast in a location other than their final position.

PRESSURE PRESERVATIVE TREATED WOOD: Wood impregnated under pressure with a chemical formulation which is toxic to fungi, insects, borers, and other wood-destroying organisms.

PRESTRESSED CONCRETE: Reinforced concrete in which internal stresses and deformations are initially introduced, of such magnitude and distribution that the subsequent stresses and deformations resulting from dead and live loads are counteracted to a desired degree.

PROCTOR DENSITY: The optimum unit weight, determined in accordance with ASTM Standard D698, of a soil.[14]

RANDOM INCIDENT (SOUND): Sound waves that strike a receiver randomly from all angles of incidence. Such waves are common in a diffuse sound field.

REMEL: Reference Energy Mean Emission Level.

RETAINING WALL: A wall built to hold back earth or water.

RESPONSIBLE ORGANIZATION: Government transportation agency, emergency response unit, fire department, police department, etc.

RIGHT-OF-WAY (ROW): The entire strip or area of land used for highway purposes.

SABINE ABSORPTION COEFFICIENT (α_{Sab}): Absorption coefficient obtained in a reverberation room by measuring the time rate of decay of the sound energy density with and without a patch of the sound-absorbing material under test laid on the floor. These measurements are performed in accordance with the American Society of Testing and Materials (ASTM) Standard C 423-90a.[15]

SHEAR SLIP CIRCLE: The arc defined by where the soil will typically fail behind a fill or retained fill section.

SHOULDER: The part of a roadway contiguous to the traffic lanes for accommodating stopped vehicles, bikeways, and cleared snow.

SLOPE HINGE POINT: The point at which the highway shoulder's slope intersects the top of the embankment for a highway located on fill (see Figure 2).

SOFT GROUND: Any highly absorptive surface in which the phase of the sound energy is changed upon reflection; examples include terrain covered with dense vegetation or freshly fallen snow. (Note: at grazing angles greater than 20 degrees, which can commonly occur at short ranges, or in the case of elevated sources, soft ground becomes a good reflector and can be considered acoustically hard ground.)

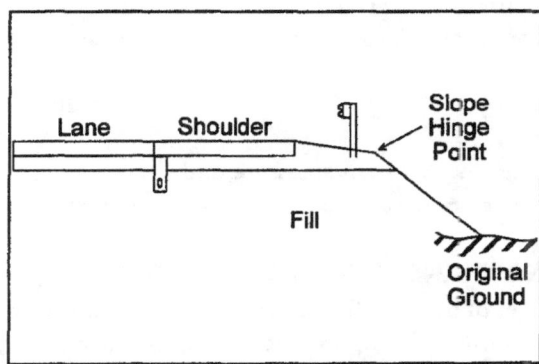

Figure 2. Example of slope hinge point.

SOUND ABSORPTION COEFFICIENT (α): (See also Sabine Absorption Coefficient.) The ratio of the sound energy, as a function of frequency, absorbed by a surface, and the sound energy incident upon that surface.

SOUND ENERGY: See Acoustic energy.

SOUND EXPOSURE LEVEL (SEL, denoted by the symbol, L_{AE}): Over a stated time interval, T (where $T=t_2-t_1$), ten times the base-10 logarithm of the ratio of a given time integral of squared instantaneous A-weighted sound pressure, and the product of the reference sound pressure of 20 µPa, the threshold of human hearing, and the reference duration of 1 sec. The time interval, T, must be long enough to include a majority of the sound source's acoustic energy. As a minimum, this interval should encompass the 10 dB down points (see Figure 3).

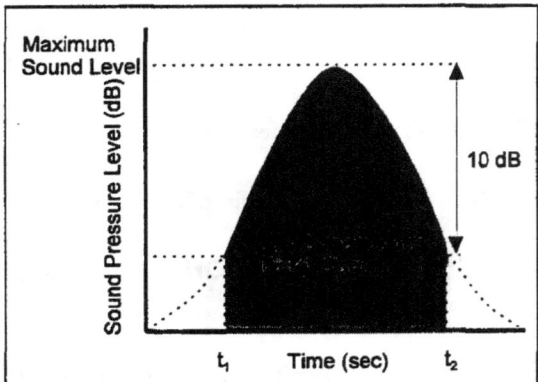

Figure 3. Graphical representation of L_{AE}.

In addition, L_{AE} is related to L_{AeqT} by the following equation:

$$L_{AE} = L_{AeqT} + 10*\log_{10}(t_2-t_1) \qquad (dB)$$

where L_{AeqT} = Equivalent sound level in dB (see definition above).

SOUND PRESSURE: The root-mean-square of the instantaneous sound pressures during a specified time interval in a stated frequency band.

SOUND PRESSURE LEVEL (SPL): Ten times the base-10 logarithm of the square of the ratio of the mean-square sound pressure, in a stated frequency band (often weighted), and the reference mean-square sound pressure of 20 µPa, the threshold of human hearing.

$$SPL = 10*\log_{10}(p^2/p_{ref}^2) \qquad (dB)$$

where:
p = mean-square sound pressure; and
p_{ref} = reference mean-square sound pressure of 20 µPa.

SOUND TRANSMISSION CLASS (STC): A single-number rating used to compare the sound insulation properties of barriers. STC is derived by fitting a reference rating curve to the sound transmission loss (TL) values measured for the 16 contiguous one-third octave frequency bands with nominal mid-band frequencies of 125 Hz to 4000 Hz inclusive, by a standard method. The reference rating curve is fitted to the 16 measured TL values such that the sum of deficiencies (TL values less than the reference rating curve), does not exceed 32 dB, and no single deficiency is greater than 8 dB. The STC value is the numerical value of the reference contour at 500 Hz.

SPALLING: Separation and removal of a portion of the surface concrete.

SPECTRUM: A signal's resolution expressed in component frequencies or fractional octave bands.

SPLIT: A lengthwise separation of the wood, which usually extends from surface to surface, due to the tearing apart of the wood cells.

SPREAD FOOTINGS: Large, horizontal, reinforced concrete slabs (see Figure 4), which transfer structure loads directly to the underlying soil, rock, or engineered soil. They are typically placed or poured just below the ground line. Noise barrier posts are either attached to or embedded into the center of these slabs. These footings support the lateral and dead loads for the noise barrier system.

STRUCTURE: In reference to barriers on structure, includes retaining walls, bridges, culverts, and concrete drainage channels.

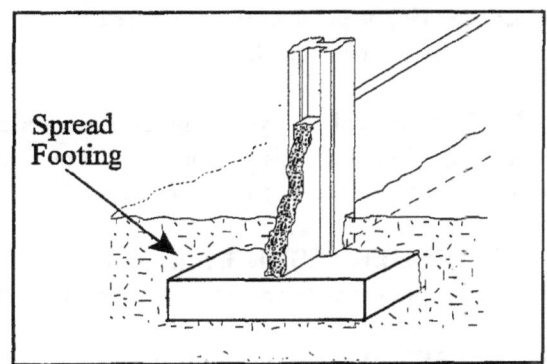

Figure 4. Example of spread footing.

SUBDRAIN: A pipe, perforated or non-perforated, which is placed in locations for the purpose of collecting sub-surface water and conveying it to a suitable outlet.

SWALE: A shallow ditch with gently sloping sides.

SWEEP: The deviation of a piece of lumber from straightness.

TEN-PERCENTILE EXCEEDED SOUND LEVEL: The sound level exceeded 10 percent of a specific time period. For example, from a 50-sample measurement period, the fifth (10% of 50 samples) highest sound level is the 10-percentile exceeded sound level. Other similar descriptors include L_{50} (the sound level exceeded 50 percent of a specific time period), L_{90} (the sound level exceeded 90 percent of a specific time period), etc.

THROUGH CHECK: A check which extends from surface to surface of the wood and usually through the center of the pith.

TRANSMISSION LOSS (TL): The loss in sound energy, expressed in decibels, as sound passes through a barrier or a wall. Measurements to determine a barrier's TL should be made in accordance with ASTM Recommended Practice E413-87.[16] TL is determined as follows:

$$TL = 10\log_{10}[10^{(SPL_s/10)}/10^{(SPL_r/10)}] \quad (dB)$$

where: SPL_s is the sound pressure level (see Section 3.1) on the source side of the barrier; and
SPL_r is the sound pressure level on the receiver side of the barrier.

UTILITIES: Transmission and distribution lines, pipes, cables, and other associated equipment used for public services, including, but not limited to, electric transmission and distribution, lighting, heating, gas, oil, water, sewage, and telephone.

WATER TABLE: The upper surface of the zone of saturation of the soil where the ground water is not confined by an overlying impermeable formation.

WAVELENGTH: The perpendicular distance between two wave fronts in which the displacements have a difference in phase of one complete period.

WEATHERING STEEL: A type of finish on steel panels and structural members which allows the surface to rust at a controlled rate. This type of surface is purported to be self protecting and normally does not require any other type of coating for protection.

WEATHEROMETER TESTING: A test procedure used to determine the effects of salt spray and fog, ultraviolet light, and severe temperature changes on a specific type of material or coating.

WIND CLASSES: Near-ground wind effects are generally separated into three wind classes: upwind, calm, and downwind. When sound is propagating during upwind conditions, sound waves tend to refract upward away from the ground, which may result in a decrease in sound levels at a receiver. When sound is propagating during downwind conditions, sound waves tend to refract downward towards the ground, which may result in an increase in sound levels at a receiver.

3. ACOUSTICAL CONSIDERATIONS

This section describes the acoustical considerations associated with highway **noise barrier*** design, beginning with a brief technical discussion on the fundamentals of highway traffic noise.

3.1 Characteristics of Sound

Highway traffic noise originates primarily from three discrete sources: truck exhaust stacks, vehicle engines, and tires interacting with the pavement. These sources each produce **sound energy** that, in turn, translates into tiny fluctuations in atmospheric pressure as the sources move and vibrate. These sound pressure fluctuations are most commonly expressed as **sound pressure** and measured in units of micro Newtons per square meter ($\mu N/m^2$), or micro Pascals (μPa). Typical sound pressure amplitudes can range from 20 to 200 million μPa. Because of this wide range, sound pressure is measured on a logarithmic scale known as the **decibel** (dB) scale. On this scale, a value of 0 dB is equal to a sound pressure level (SPL) of 20 μPa and corresponds to the threshold of hearing for most humans. A value of 140 dB is equal to an SPL of 200 million μPa, which is the threshold of pain for most humans.[17]

The following figure shows a scale relating various sounds encountered in daily life and their approximate decibel values:

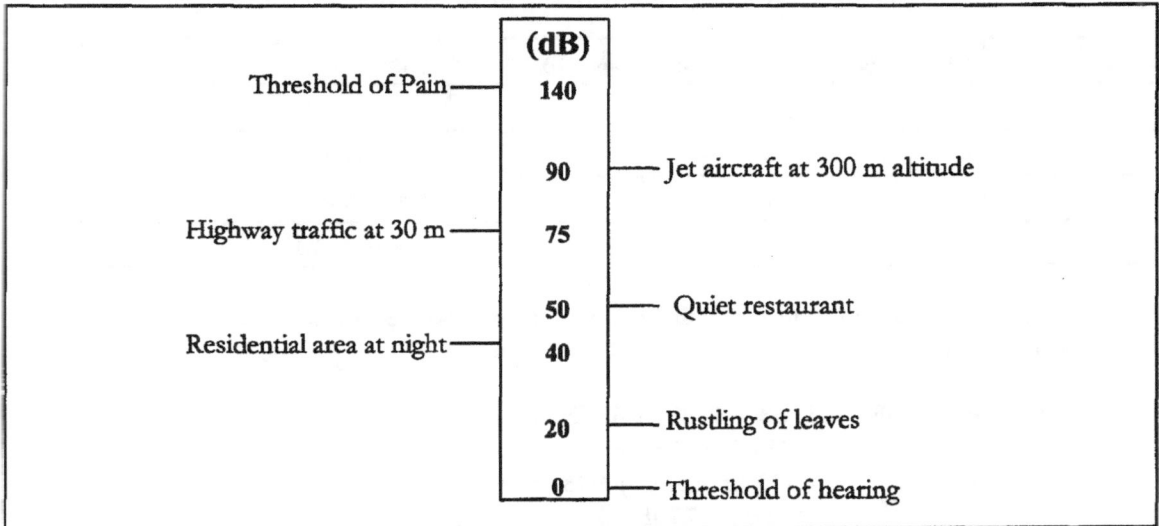

Figure 5. Decibel scale.

---

* As previously noted, all terms defined in the Terminology section (Section 2) are highlighted when they first appear in the main body text of this document.

To express a sound's energy, or sound pressure in terms of SPL, or dB, the following equation is used:

$$SPL = 10 \cdot \log_{10}(p/p_{ref})^2 \quad dB$$

where: p is the sound pressure; and
p_{ref} is the reference sound pressure of 20 µPa

Conversely, sound energy is related to SPL as follows:

$$(p/p_{ref})^2 = 10^{(SPL/10)}$$

The above relationships are important in understanding the way decibel levels are combined; i.e., added or subtracted. That is, because decibels are expressed on a logarithmic scale, they cannot be combined by simple addition. For example, if a single vehicle pass-by produces an SPL of 60 dB at a distance of 15 m (50 ft) from a roadway, two identical vehicle pass-bys would not produce an SPL of 120 dB. They would, in fact, produce an SPL of 63 dB. To combine decibels, they must first be converted to energy, then added or subtracted as appropriate, and reconverted back to decibels. The following table may be used as an approximation to adding decibel levels (Note: Table approximations are within ±1 dB of the exact value).

Table 1. Decibel addition approximation.

When two decibel values differ by (dB)	Add to higher value (dB)	Example
0 to 1	3	50 + 51 = 54
2 to 3	2	62 + 65 = 67
4 to 9	1	65 + 71 = 72
10 or more	0	55 + 65 = 65

The above table can also be used to approximate the sum of more than two decibel values. First, rank the values from low to high, then add the values two at a time. For example:

60 dB + 60 dB + 65 dB + 75 dB = (60 dB + 60 dB) + 65 dB + 75 dB
= 63 dB + 65 dB + 75 dB
= (63 dB + 65 dB) + 75 dB
= 67 dB + 75 dB
= 76 dB

In the above example, the exact value would be computed as follows:

60 dB + 60 dB + 65 dB + 75 dB = $10 \cdot \log_{10} [10^{(60/10)} + 10^{(60/10)} + 10^{(65/10)} + 10^{(75/10)}]$
= 75.66 dB

The next characteristic of sound is its **amplitude**, or loudness. As stated earlier, sound sources produce sound energy that, in turn, translates into tiny fluctuations in atmospheric pressure as the sources move and vibrate. As the sources move and vibrate, surrounding atoms, or molecules, are temporarily displaced from their normal configurations thus forming a disturbance that moves away from the sound source in waves that pulsate out at equal intervals. For simplicity, the outward propagating waves can be approximated by the trigonometric sine function (see Figure 6). The "height" of the sine wave from peak to peak is referred to as its amplitude. The length between wave repetitions is referred to as the **wavelength** (λ). The amplitude determines the strength, or loudness, of the wave.

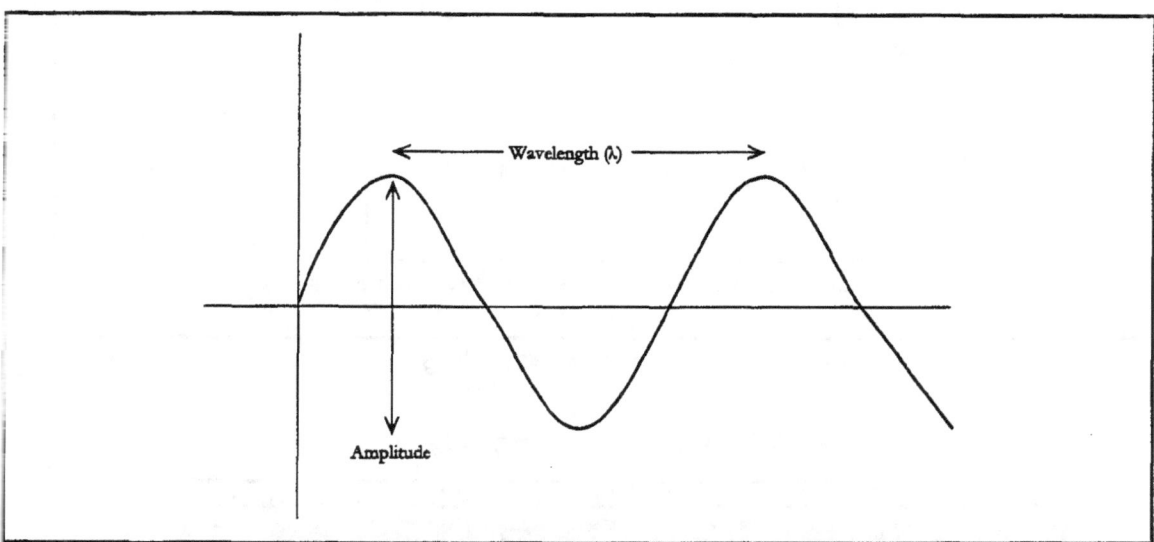

Figure 6. Sound wave amplitude and wavelength.

Finally, another characteristic of sound is its **frequency**, or tonality, measured in Hertz (Hz), or cycles per second. Frequency is defined as the number of cycles of repetition per second, or the number of wavelengths that have passed by a stationary point in one second.

Most humans can hear in a range from 20 Hz to 20,000 Hz. However, the human ear is not equally sensitive to all frequencies. To account for this, most transportation-related noise, including highway traffic noise, is measured using an **"A-weighted"** response network. A-weighting emphasizes sounds between 1,000 Hz and 6,300 Hz, and de-emphasizes sounds above and below that range to simulate the response of the human ear. Figure 7 presents the A-weighting curve as a function of frequency. Table 2 presents the curve in tabular form for one-third octave band frequencies from 20 to 20,000 Hz. Sound levels measured using the A-weighting network are expressed in units of dB(A).[12]

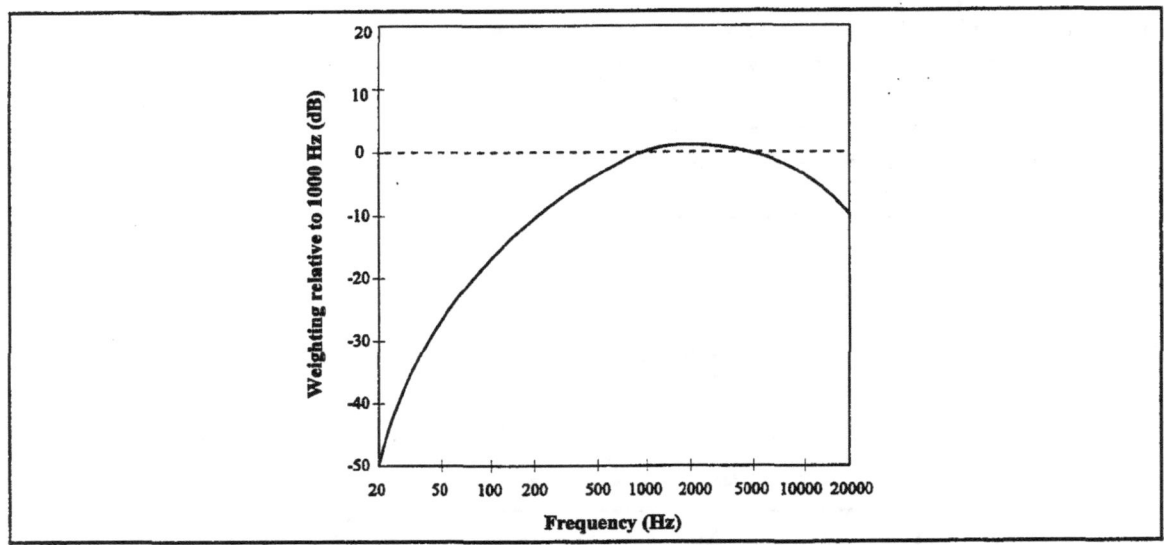

Figure 7. Frequency A-weighting.

Table 2. Frequency A-weighting.

One-Third Octave-Band Center Frequency (Hz)	Response, re: 1000 Hz	One-Third Octave-Band Center Frequency (Hz)	Response, re: 1000 Hz
20	-50.5	800	-0.8
25	-44.7	1000	0.0
31.5	-39.4	1250	0.6
40	-34.6	1600	1.0
50	-30.2	2000	1.2
63	-26.2	2500	1.3
80	-22.5	3150	1.2
100	-19.1	4000	1.0
125	-16.1	5000	0.5
160	-13.4	6300	-0.1
200	-10.9	8000	-1.1
250	-8.6	10000	-2.5
315	-6.6	12500	-4.3
400	-4.8	16000	-6.6
500	-3.2	20000	-9.3
630	-1.9		

3.2 Noise Descriptors

Noise descriptors provide a mechanism for describing sound for different applications. As stated previously, sound levels measured for highway traffic noise use an A-weighting filter to more accurately simulate the response of the human ear. An A-weighted sound level is denoted by the symbol, L_A. Other noise descriptors include the **maximum sound level** (MXFA or MXSA, denoted by the symbol, L_{AFmx} or L_{ASmx}), the **equivalent sound level** for a one-hour period (1HEQ, denoted by the symbol, L_{Aeq1h}), the **sound exposure level** (SEL, denoted by the symbol, L_{AE}), the **day-night average sound level** (DNL, denoted by the symbol, L_{dn}), the **community noise equivalent level** (CNEL, denoted by the symbol, L_{den}), and the **ten-percentile exceeded sound level** (denoted by the symbol, L_{10}). These descriptors are defined in Section 2.

For highway traffic noise, the L_{Aeq1h} are most often used to describe continuous sounds, such as relatively dense highway traffic. The L_{ASmx} and L_{AE} may be used to describe single events, such as an individual vehicle pass-by. Note that the L_{AE} is more commonly used to describe an aircraft overflight. The L_{dn} and the L_{den} may be used to describe long-term noise environments (typically 24 hours or more).

3.3 Sound Propagation

The sound that reaches a receiver is affected by many factors. These factors include:[18]

- **Divergence** (Section 3.3.1);
- **Ground effect** (Section 3.3.2);
- Meteorological effects (Section 3.3.3); and
- Shielding by natural and man-made structures; e.g., trees and buildings (Section 3.3.4). Note: Shielding by man-made noise barriers will be discussed separately in Section 3.4.

3.3.1 Divergence. Divergence is referred to as the spreading of sound waves from a sound source in a **free field** environment. In the case of highway traffic noise, two types of divergence are common, spherical and cylindrical. Spherical divergence is that which would occur for sound emanating from a **point source**; e.g., a single vehicle pass-by. The attenuation of sound over distance due to spherical spreading is illustrated using the following equation:

$$L_2 = L_1 + 20*\log_{10}(d_1/d_2) \qquad dB(A)$$

where: L_1 is the sound level at distance d_1; and
L_2 is the sound level at distance d_2

Thus, with this equation, it can be shown that sound levels measured from a point source decrease at a rate of 6 dB(A) per doubling of distance. For example, if the sound level from a point source at 15 m was 90 dB(A), at 30 m it would be 84 dB(A) due to divergence; i.e., $90 + 20*\log_{10}(15/30)$.

Cylindrical divergence is that which would occur for sound emanating from a **line source**, or many point sources sufficiently close to be effectively considered as a line source; e.g., a continuous stream of roadway traffic. The attenuation of sound over distance due to cylindrical spreading is illustrated using the following equation:

$$L_2 = L_1 + 10*\log_{10}(d_1/d_2) \qquad dB(A)$$

With this equation, it can be shown that sound levels measured from a line source decrease at a rate of 3 dB(A) per doubling of distance. For example, if the sound level from a line source at 15 m was 90 dB(A), at 30 m it would be 87 dB(A) due to divergence; i.e., 90 + 10*$\log_{10}$(15/30).[19]

3.3.2 Ground Effect.
Ground effect refers to the change in sound level, either positive or negative, due to intervening ground between source and receiver. Ground effect is a relatively complex acoustic phenomenon, which is a function of ground characteristics, source-to-receiver geometry, and the spectral characteristics of the source. Ground types are typically characterized as acoustically hard or acoustically soft. **Hard ground** refers to any highly reflective surface in which the phase of the sound energy is essentially preserved upon reflection; examples include water, asphalt, and concrete. For practical highway applications, measurements have shown a 1 to 2 dBA increase for the first and second row residences adjacent to the highway. **Soft ground** refers to any highly absorptive surface in which the phase of the sound energy is changed upon reflection; examples include terrain covered with dense vegetation or freshly fallen snow.[19] An acoustically soft ground can cause a significant broadband attenuation (except at low frequencies).

A commonly used rule-of-thumb is that: (1) for propagation over hard ground, the ground effect is neglected; and (2) for propagation over acoustically soft ground, for each doubling of distance the soft ground effect attenuates the sound pressure level at the receiver by an additional 1.5 dB(A). This extra attenuation applies to only incident angles of 20 degrees or less. For greater angles, the ground becomes a good reflector and can be considered acoustically hard. Keep in mind that these relationships are quite empirical but tend to break down for distances greater than about 30.5 to 61 m (100 to 200 ft). For a more detailed discussion of ground effects, the reader is directed to References 20 and 21.

3.3.3 Atmospheric Effects.
Atmospheric effects refer to: (1) atmospheric absorption; i.e., the sound absorption by air and water vapor; (2) atmospheric refraction; i.e., the sound refraction caused by temperature and wind gradients; and (3) air turbulence.[18] It is recommended that when atmospherics are of potential concern, high-precision meteorological measurement equipment should be used to record continuous temperature, relative humidity, and wind data.

- Atmospheric absorption: Atmospheric absorption is a function of the frequency of the sound, the temperature, the humidity, and the atmospheric pressure between the source and the receiver.[22,23] Over distances greater than 30 m (100 ft), the attenuation due to atmospheric absorption can substantially reduce sound levels, especially at high frequencies (above 5000 Hz).

- Atmospheric refraction: Atmospheric refraction is the bending of sound waves due to wind and temperature gradients. Near-ground wind effects are, typically, the most substantial contributor to sound refraction. Upwind conditions tend to refract sound waves away from the ground resulting in a decrease in sound levels at a receiver. Conversely, downwind conditions tend to refract sound waves towards the ground resulting in an increase in sound levels at a receiver. Studies have shown measured sound levels to be affected by up to 7 dB(A) as a result of wind refraction within just 100 m from the centerline of the roadway.[24,25] It is generally recommended that highway traffic noise measurements be performed when the recorded wind speed is no greater than 5 m/s (11 mph) to minimize the effects of wind. Further, measurements should not be performed in conditions where strong winds with small vector components exist in the direction of propagation. Readers may refer to Reference 18 for more information on performing highway-related noise measurements.

Temperature effects can also contribute to sound refraction. During daytime weather conditions, when the air is warmer closer to the ground (temperature decreases with height), sound waves tend to refract upward away from the ground (temperature lapse). This may result in a decrease in sound levels at a receiver. Conversely, when the air close to the ground cools during nighttime weather conditions (temperature increases with height), sound waves tend to refract downward towards the ground (temperature inversion). This may result in an increase in sound levels at a receiver.[26] Generally, refraction effects due to temperature do not exert a substantial influence on sound levels within 61 m (200 ft) of the roadway.[24]

- Air turbulence: Although, its effects on sound levels are more unpredictable than other atmospheric effects, in certain cases air turbulence has shown an even greater effect on noise levels than atmospheric refraction within 122 m (400 ft) from a roadway.[25] As stated earlier, it is generally recommended that highway traffic noise measurements be performed when the recorded wind speed is no greater than 5 m/s to ensure minimal effects of wind. Further, measurements should not be performed in conditions where strong winds with small vector components exist in the direction of propagation. Readers may refer to Reference 18 for more information on performing highway-related noise measurements.

3.3.4 Shielding by Natural and Man-Made Structures.
In this section, shielding by structures, such as trees and buildings, will be discussed. The amount of attenuation provided by these structures is determined by their size and density, and the frequencies of the sound levels. Note that shielding by noise barriers will be discussed separately in Section 3.4.

Shielding by trees and other such vegetation typically only have an "out of sight, out of mind" effect. That is, the perception of highway traffic noise impact tends to decrease when vegetation blocks the **line-of-sight** to nearby residents (i.e., "out of sight, out of mind"). However, for vegetation to provide a substantial, or even noticeable, noise reduction, the vegetation area must be at least 5 m (15 ft) in height, 30 m (100 ft) wide and dense enough to completely obstruct the line-of-sight between the source and the receiver. This size of vegetation area may provide up to 5 dB(A) of noise reduction. Taller, wider, and denser areas of vegetation may provide even greater noise reduction. The maximum reduction that can be achieved is approximately 10 dB(A).[5,23]

Shielding by a building is similar to the shielding effects of a short (lengthwise) barrier. Building rows can act as longer barriers keeping in mind that the gaps between buildings will leak sound through to the receiver. Generally, assuming an **at-grade** building row with a building-to-gap ratio of 40 percent to 60 percent, the noise reduction due to this row is approximately 3 dB(A). Further, for each additional building row, another 1.5 dB(A) noise reduction may be considered typical.[3,27] For situations where the buildings in a building row occupy less than 20 percent of the row area, unless the receiver is directly behind a building, minimal, or no, attenuation should be assumed. For situations where the buildings in a building row occupy greater than 80 percent of the row area, it may be assumed that the leakage of sound due to gaps is minimal. In this case, noise attenuation may be determined by treating the building row as a noise barrier, which is discussed in Section 3.4.

3.4 Noise Barrier Basics

As shown in Figure 8, noise barriers reduce the sound which enters a community from a busy highway by either absorbing it (see Section 3.4.1), transmitting it (see Section 3.4.2), reflecting it back across the highway (see Section 3.5.4), or forcing it to take a longer path. This longer path is referred to as the **diffracted** path.

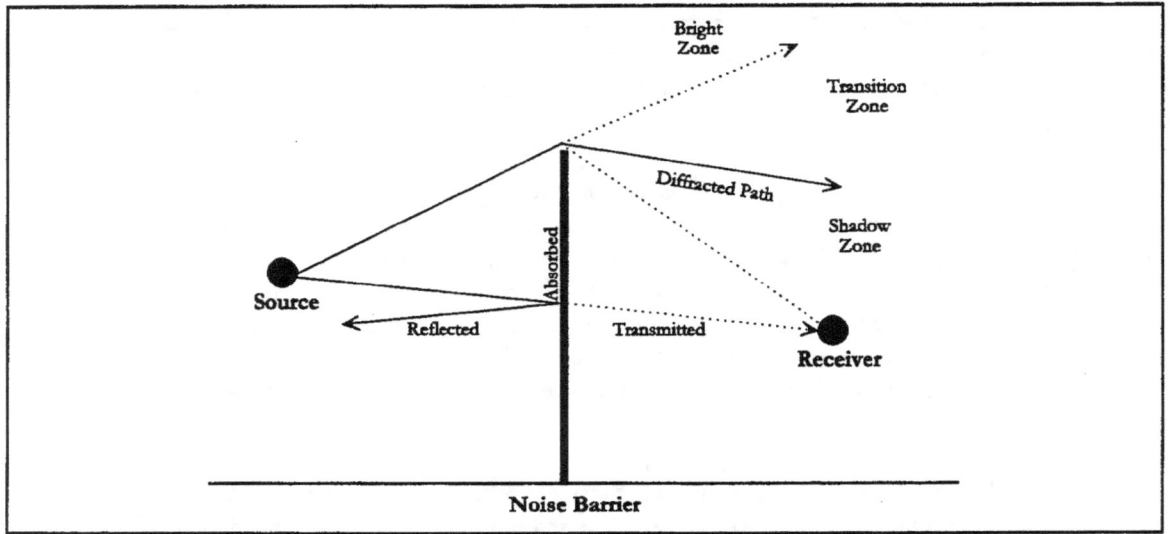

Figure 8. Barrier absorption, transmission, reflection, and diffraction.

Diffraction, or the bending of sound waves around an obstacle, can occur both at the top of the barrier and around the ends. This bending occurs much like other wave phenomena, such as light and water waves. Due to the nature of sound waves, diffraction does not bend all frequencies uniformly. Higher frequencies (shorter wavelengths) are diffracted to a lesser degree; while lower frequencies (longer wavelengths) are diffracted deeper into the "shadow" zone behind the barrier. As a result, a barrier is, generally, more effective in attenuating the higher frequencies as compared with the lower frequencies (see Figure 9).[18]

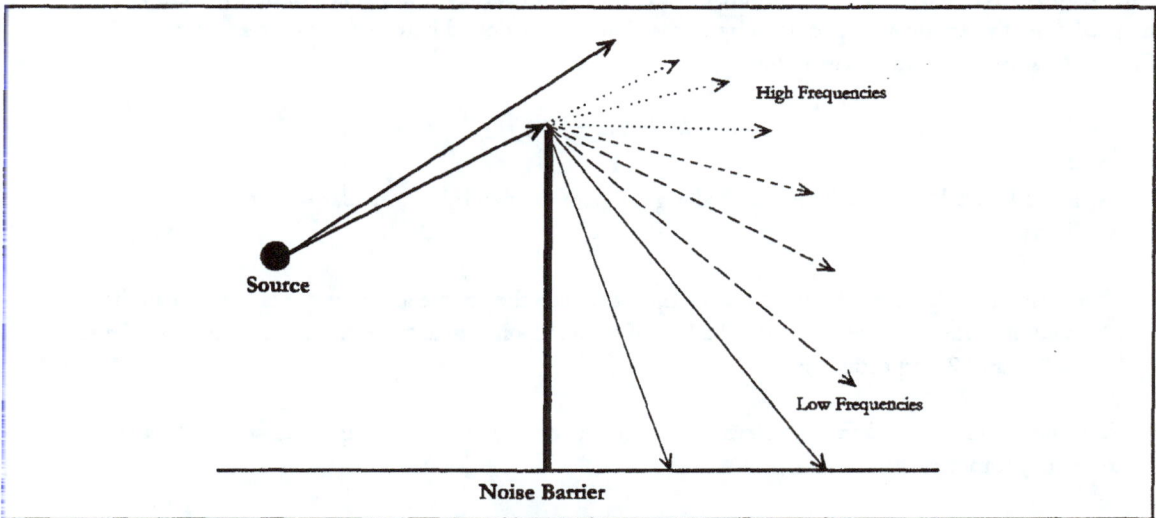

Figure 9. Barrier diffraction.

An important aspect of diffraction is the path length difference (δ) between the diffracted path from source over the top of the barrier to the receiver, and the direct path from source to receiver as if the barrier were not present (see Figure 10).

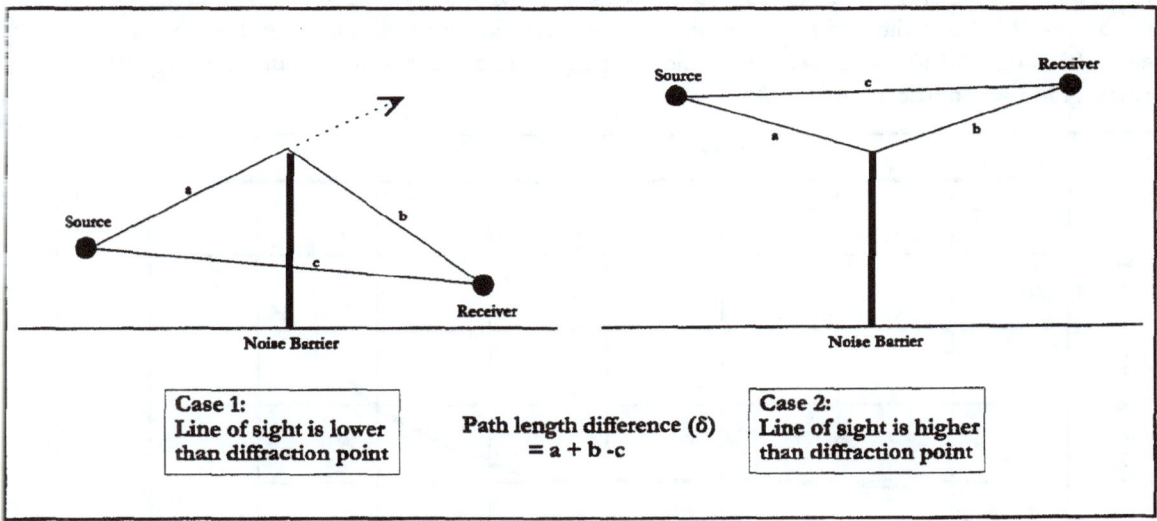

Figure 10. Path length difference.

The path length difference is used to compute the **Fresnel Number** (N_0), which is a dimensionless value used in predicting the attenuation provided by a noise barrier positioned between a source and a receiver. The Fresnel Number is computed as follows:

$$N_0 = \pm 2(\delta_0/\lambda) = \pm 2(f\,\delta_0/c)$$

where: N_0 is the Fresnel Number determined along the path defined by a particular source-barrier-receiver geometry;

$\pm$ is positive in the case where the line of sight between the source and receiver is lower than the diffraction point and negative when the line of sight is higher than the diffraction point (see Figure 9 - Case 1 and 2, respectively);[28]

δ_0 is the path length difference determined along the path defined by a particular source-barrier-receiver geometry;

λ is the wavelength of the sound radiated by the source;

f is the frequency of the sound radiated by the source; and

c is the speed of sound.

Note the relationship between the variables in the above equation. If the path length difference increases, the Fresnel number and, thus, barrier attenuation increases. If the frequency increases, barrier attenuation increases as well. Figure 11 shows the relationship between barrier attenuation and Fresnel Number for a frequency of 550 Hz. A 550 Hz frequency is considered fairly representative for computing barrier attenuation of highway traffic noise.[29]

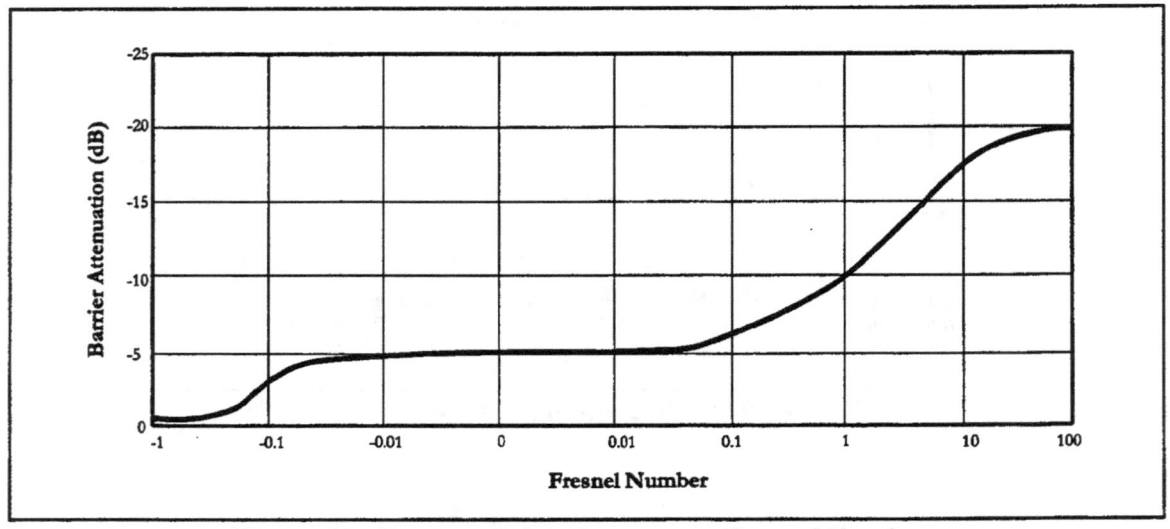

Figure 11. Barrier attenuation versus Fresnel Number.

3.4.1 Barrier Absorption.
The amount of incident sound that a barrier absorbs is typically expressed in terms of its **Noise Reduction Coefficient** (NRC). NRC is defined as the arithmetic average of the **Sabine absorption coefficients**, α_{sab}, at 250 Hz, 500 Hz, 1000 Hz, and 2000 Hz:

$$\text{NRC} = \tfrac{1}{4} \times (\alpha_{250} + \alpha_{500} + \alpha_{1000} + \alpha_{2000})$$

NRC values can range from zero to one; where zero indicates the barrier will reflect all the sound incident upon it (see also Section 3.5.4), and one indicates the barrier will absorb all the sound incident upon it. A typical NRC for an absorptive barrier ranges from 0.6 to 0.9.[19]

Measurements to determine the α_{sab} of a barrier facade should be made in accordance with the ASTM Recommended Practice C384 (Impedance Tube Method) or C423 (Reverberation Room Method). The Impedance Tube Method can be used to measure the sound absorption of **normal incident** sound on a small sample of a material.[15,30] The Reverberation Room Method (see Figure 12*) is used to measure the sound absorption of **random incident** sound on a larger sample of a material. Most barrier manufacturers prefer to use the Reverberation Room Method because of its lack of constraints on sample size. However, for this Method, the sample size chosen and method and angle of mounting may have substantial effects on the determined absorption coefficients. These concerns are further addressed in Reference 31.

Figure 12. Barrier absorption: Reverberation Room Method (data base #2553).

3.4.2 Barrier Sound Transmission.
The amount of incident sound that a barrier transmits can be described by its sound **Transmission Loss** (TL). Measurements to determine a barrier's TL should be made in accordance with ASTM Recommended Practice E413-87.[16] TL is determined as follows:

$$\text{TL} = 10\log_{10}[10^{(\text{SPL}_s/10)} / 10^{(\text{SPL}_r/10)}] \qquad \text{dB(A)}$$

where: SPL_s is the sound pressure level (see Section 3.1) on the source side of the barrier; and
SPL_r is the sound pressure level on the receiver side of the barrier.

For highway noise barriers, any sound that is transmitted through the barrier can be effectively neglected since it will be at such a low level relative to the diffracted sound; i.e., the sound transmitted will typically be at least 20 dB(A) below that which is diffracted. That is, if a sound level of 100 dB(A) is incident upon a

---

* All photographs contained in this document are catalogued in a data base (each photo has a data base number) which can be found on the FHWA Highway Noise Barrier Design CD-ROM. The CD-ROM contains these and additional related photographs with detailed information including barrier material, location, special features, etc.

barrier and only 1 dB(A) is transmitted, i.e, 1 percent of the incident sound's energy, then a TL of 20 dB(A) is achieved.

As a rule of thumb, any material weighing 20 kg/m^2 (4 lbs/ft^2) or more has a transmission loss of at least 20 dB(A). Such material would be adequate for a noise reduction of at least 10 dB(A) due to diffraction. Note that a weight of 20 kg/m^2 (4 lbs/ft^2) can be attained by lighter and thicker, or heavier and thinner materials. The greater the density of the material, the thinner the material may be. TL also depends on the stiffness of the barrier material and frequency of the source.[18]

In most cases, the maximum noise reduction that can be achieved by a barrier is 20 dB(A) for thin walls and 23 dB(A) for berms. Therefore, a material that has a TL of at least 25 dB(A) or greater is desired and would always be adequate for a noise barrier. The following table gives approximate TL values for some common materials, tested for typical A-weighted highway traffic frequency spectra. They may be used as a rough guide in acoustical design of noise barriers. For accurate values, consult material test reports by accredited laboratories.

Table 3. Approximate sound transmission loss values for common materials.

Material	Thickness mm (inches)	Weight kg/m^2 (lbs/ft^2)	Transmission Loss (dB(A))
Concrete Block, 200mm x 200mm x 405 (8" x 8" x 16") light weight	200mm (8")	151 (31)	34
Dense Concrete	100mm (4")	244 (50)	40
Light Concrete	150mm (6")	244 (50)	39
Light Concrete	100mm (4")	161 (33)	36
Steel, 18 ga	1.27mm (.0.050")	10 (2.00)	25
Steel, 20 ga	0.95mm (0.0375")	7.3 (1.50)	22
Steel, 22 ga	0.79mm (0.0312")	6.1 (1.25)	20
Steel, 24 ga	0.64mm (0.025")	4.9 (1.00)	18
Aluminum, Sheet	1.59mm (0.0625")	4.4 (0.9)	23
Aluminum, Sheet	3.18mm (0.125")	8.8 (1.8)	25
Aluminum, Sheet	6.35mm (0.25")	17.1 (3.5)	27
Wood, Fir	12mm (0.5")	8.3 (1.7)	18
Wood, Fir	25mm (1.0")	16.1 (3.3)	21
Wood, Fir	50mm (2.0")	32.7 (6.7)	24
Plywood	12mm (0.5")	8.3 (1.7)	20
Plywood	25mm (1.0")	16.1 (3.3)	23
Glass, Safety	3.18mm (0.125")	7.8 (1.6)	22
Plexiglass	6mm (0.25")	7.3 (1.5)	22

The above table assumes no openings or gaps in the barrier material. Some materials, such as wood, however, are prone to develop openings or gaps due to shrinkage, warping, splitting, or weathering. Treatments to reduce/eliminate noise leakage for wood barrier systems are discussed in Section 5.4.1. Noise leakage due to possible gaps in the horizontal joints between **panels** in a post and panel "stacked panel" barrier system (see Section 4.1.2.1) should also be given careful consideration. Finally, some barrier systems are designed with small openings at the base of the barrier to carry water, which would otherwise pond on one side of the barrier, through the barrier. Two important considerations associated with these openings are: (1) ensure that the opening is small [the effect of a continuous gap of up to 20 cm (7.8 in) at the base of a noise barrier is usually within 1 dB(A)],[32] and (2) ensure that proper protection in the form of grates or bars is provided to restrict entry by small animals (cats, small dogs, etc.). Drainage considerations are also discussed in Section 7.1.2 and 7.1.3.

It should be noted that there are other ratings used to express a material's sound transmission characteristics. One rating in common use is the **Sound Transmission Class** (STC). STC is a single-number rating derived by fitting a reference rating curve to the TL values measured for the one-third octave frequency bands between 125 Hz and 4000 Hz. The reference rating curve is fitted to the TL values such that the sum of deficiencies (TL values less than the reference rating curve), does not exceed 32 dB, and no single deficiency is greater than 8 dB. The STC value is the TL value of the reference **contour** at 500 Hz. The disadvantage to using the STC rating scheme is that it is designed to rate noise reductions in frequencies of normal speech and office areas, and not for the lower frequencies of highway traffic noise. For frequencies of traffic noise, the STC is typically 5 to 10 dB(A) greater than the TL and, thus, should only be used as rough guide.

3.5 Barrier-Design Acoustical Considerations

This section describes the various acoustical considerations involved in actual noise barrier design. Non-acoustical design considerations will be discussed in Sections 4 to 13). The acoustical considerations include:

- Barrier design goals and insertion loss (Section 3.5.1);
- Barrier length (Section 3.5.2);
- Wall versus berm (Section 3.5.3);
- Reflective versus absorptive (Section 3.5.4);
- Other miscellaneous design considerations (Section 3.5.5).

3.5.1 Barrier Design Goals and Insertion Loss.

The first step in barrier design is to establish the design goals. Design goals may not be limited simply to noise reduction at receivers, but may also include other considerations of safety and maintenance as well. These other considerations are discussed later in Sections 4 through 13.

In this section, the acoustical design goals of noise reduction will be discussed. Acoustical design goals are usually referred to in terms of barrier **Insertion Loss** (IL). IL is defined as the sound level at a given receiver before the construction of a barrier minus the sound level at the same receiver after the construction of the barrier. The construction of a noise barrier usually results in a partial loss of soft-ground attenuation. This is due to the barrier forcing the sound to take a higher path relative to the ground plane. Therefore, barrier IL is the net effect of barrier diffraction, combined with this partial loss of soft-ground attenuation.

Typically, a 5-dB(A) IL can be expected for receivers whose line-of-sight to the roadway is just blocked by the barrier. A general rule-of-thumb is that each additional 1 m of barrier height above line-of-sight blockage will provide about 1.5 dB(A) of additional attenuation (see Figure 13).

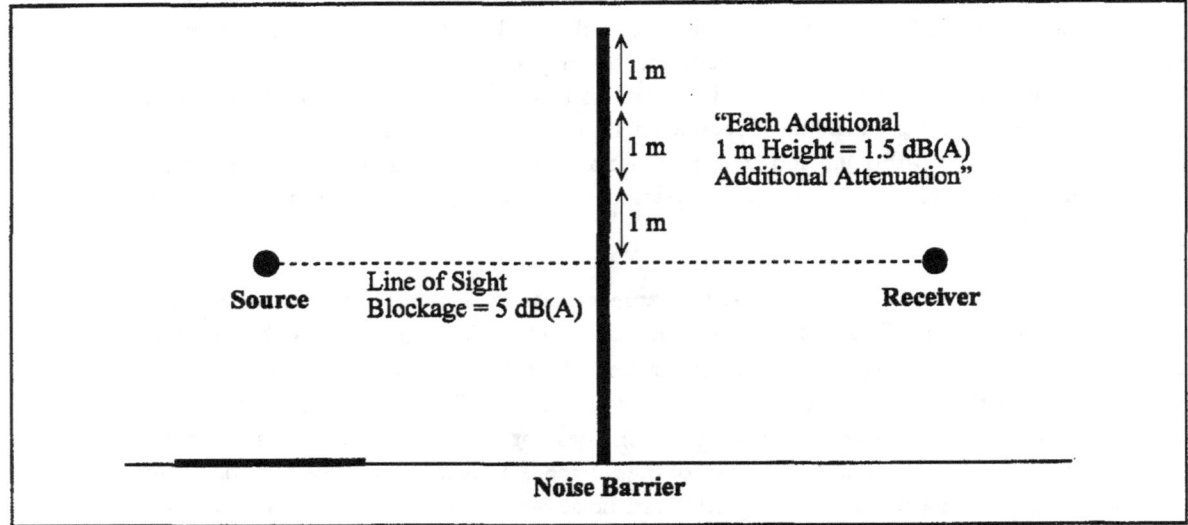

Figure 13. Line-of-sight.

Properly-designed noise barriers should attain an IL approaching 10 dB(A), which is equivalent to a perceived halving in loudness for the first row of homes directly behind the barrier. For those residents not directly behind the barrier, a noise reduction of 3 to 5 dB(A) can typically be provided, which is just slightly perceptible to the human ear. Table 4 shows the relationship between barrier IL and design feasibility.[1]

Table 4. Relationship between barrier insertion loss and design feasibility.

Barrier Insertion Loss	Design Feasibility	Reduction in Sound Energy	Relative Reduction in Loudness
5 dB(A)	Simple	68%	Readily perceptible
10 dB(A)	Attainable	90%	Half as loud
15 dB(A)	Very difficult	97%	One-third as loud
20 dB(A)	Nearly impossible	99%	One-fourth as loud

3.5.2 Barrier Length.
Noise barriers should be tall enough and long enough so that only a small portion of sound diffracts around the edges. If a barrier is not long enough, **degradations** in barrier performance of up to 5 dB(A) less than the barrier's design noise reduction may be seen for those receivers near the barrier ends. A rule-of-thumb is that a barrier should be long enough such that the distance between a receiver and a barrier end is at least four times the perpendicular distance from the receiver to the barrier along a line drawn between the receiver and the roadway (see Figure 14). Another way of looking at

this rule is that the angle subtended from the receiver to a barrier end should be at least 80 degrees, as measured from the perpendicular line from the receiver to the roadway.

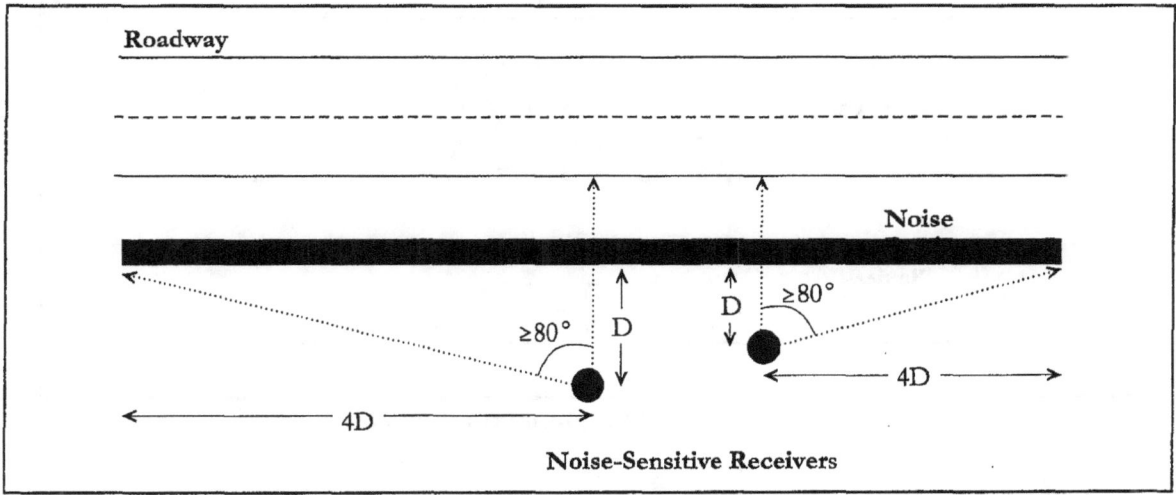

Figure 14. Barrier length.

Sometimes due to the community and roadway geometry, there is not enough available area to ensure a proper-length barrier. In those cases, highway barrier designers may decide to construct the barrier with the ends curved inward towards the community (see Figure 15).

3.5.3 Wall Versus Berm.
Highway noise barriers are typically characterized as a wall, a berm, or a combination of the two (see Figure 16). There are advantages and disadvantages to each type. The considerations that are examined in deciding whether to

Figure 15. Barrier curved inward toward the community (data base #2617).

build a wall or a berm, include available area, materials, costs, aesthetics, and community concerns. Acoustically, for a given site geometry and comparable barrier height and length, a berm barrier will typically provide an extra 1 to 3 dB(A) of attenuation. Several factors contribute to this increase. First, the flat top of a berm diffracts the sound waves twice, resulting in a longer path-length difference, a larger Fresnel number, and, thus, more attenuation. Second, the surface of a berm is, essentially, grass-covered acoustically soft earth with side slopes closer to the sound path, which provides additional attenuation. However, because a berm is wider than a wall (thus, requiring more land than a wall when constructed) and because the 1 to 3 dB(A) additional attenuation is, at best, only barely perceptible to the human ear, a berm's acoustical advantage does not necessarily guarantee its choice versus a wall.

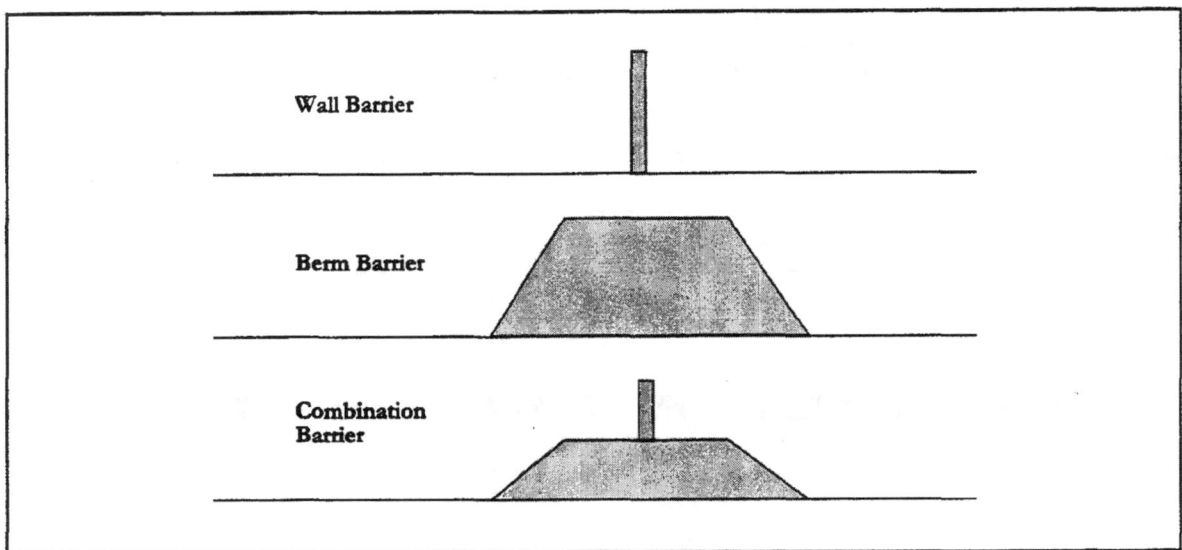

Figure 16. Wall, berm, and combination noise barriers.

3.5.4 Reflective Versus Absorptive. A barrier without any added absorptive treatment is by default reflective (see also Section 3.4.1). A reflective barrier on one side of the roadway can result in some sound energy being reflected back across the roadway to receivers on the opposite side (see Figure 17).

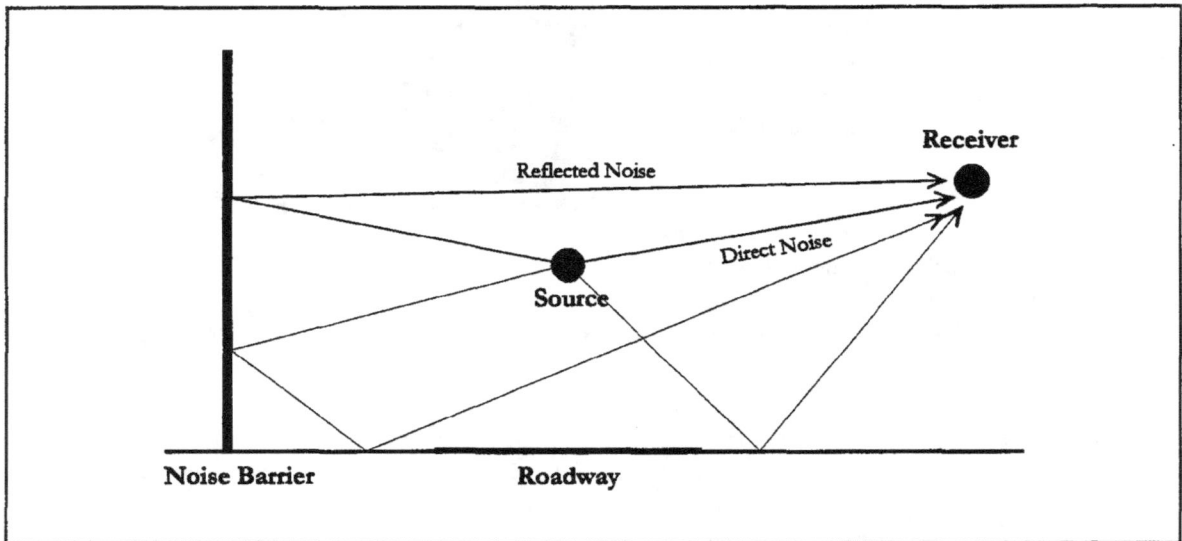

Figure 17. Reflective noise paths due to a single barrier.

It is a common phenomenon for residents to perceive a difference in sound after a barrier is installed on the opposite side of a roadway. Although theory indicates greater increases for a single reflection, practical highway measurements commonly show not greater than a 1 to 2 dB(A) increase in sound levels due to the sound reflected off the opposing barrier. While this increase may not be readily perceptible, residents on the opposite side of the roadway may perceive a change in the quality of the sound; the signature of the reflected sound may differ from that of the source due to a change in frequency content upon reflection.

Parallel barriers are two barriers which face each other on opposite sides of a roadway (see Figure 18). Sound reflected between reflective parallel barriers may cause degradations in each barrier's performance due to multiple reflections that diffract over the individual barriers. These degradations may be from 2 to as much as 6 dB(A) (see Figure 19).[19] That is, a single barrier with an insertion loss of 10 dB(A) may only realize an effective reduction of 4 to 8 dB(A) if another barrier is placed parallel to it on the opposite side of the highway.

Figure 18. Parallel noise barriers (data base #2968).

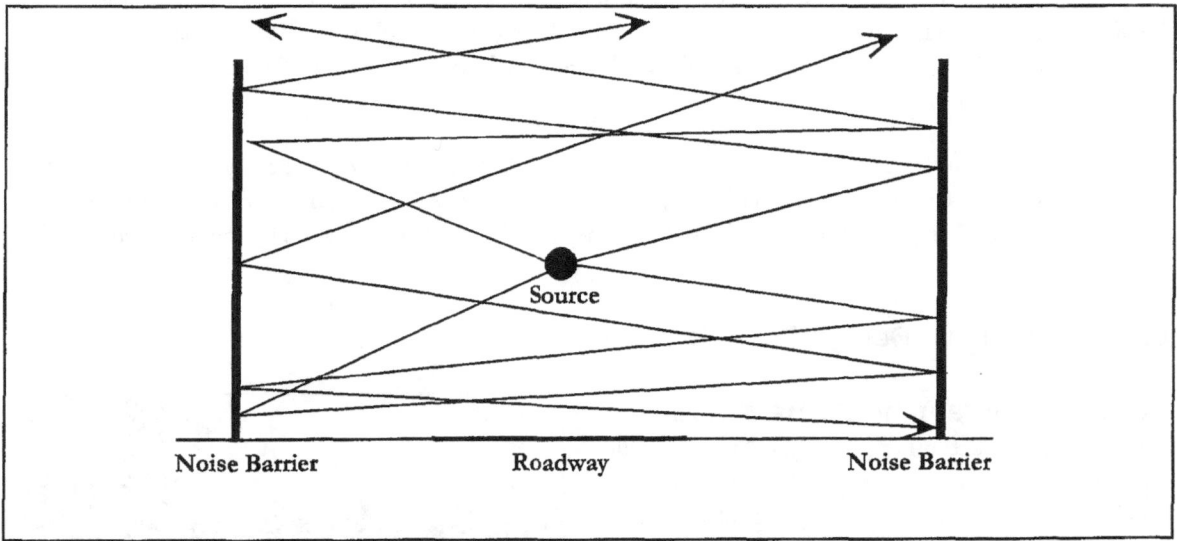

Figure 19. Reflective noise paths due to a parallel barrier.

The problems caused by both single and parallel barriers can be minimized using one or a combination of the following three methods:[19]

- For parallel barriers, ensure that the distance between the two barriers is at least 10 times their average height. A 10:1 width-to-height (w/h) ratio will result in an imperceptible degradation in performance. In recent studies, it was determined that as the w/h ratio increases, the insertion loss degradation decreases.[24,33] This decrease can be attributed to: (1) the decrease in the number of reflections between the barriers; and (2) the weakening of the reflections due to geometrical spreading and atmospheric absorption. Table 5 provides a guideline of three, general w/h ratio ranges and the corresponding barrier insertion-loss degradation (Δ_{IL}) that can be expected.

Table 5. Guideline for categorizing parallel barrier sites based on the w/h ratio.

w/h Ratio	Maximum Δ_{IL} in dB(A)	Recommendation
Less than 10:1	3 or greater	Action required to minimize degradation.
10:1 to 20:1	0 to 3	At most, degradation barely perceptible; no action required in most instances.
Greater than 20:1	No measurable degradation	No action required.

- Apply sound absorptive material on either one or both barrier facades. See also Section 3.4.1. The decision to add a sound absorptive surface should be determined by weighing benefit versus cost. That is, what noise abatement benefits can be achieved for how many residents versus the costs of the application and maintenance of the absorptive treatments?

  The answer is most important since the typical costs of noise absorptive material, whether integrated with the noise barrier at the time of barrier construction, or as a retrofit later on after the barrier is constructed, is usually $75 to $118/m² ($7 to $11/ft²). Using an average cost of $97/m² ($9/ft²) for example, for a 3.6-m (12 ft) high barrier, this would translate into an additional $0.4 million/km ($0.6 million/mi) in costs.[24,34,35,36,37]

- Tilt one or both of the barriers outward away from the road. Previous research has shown that an angle as small as 7 degrees is effective at minimizing degradations.[33] This solution, however, must consider locations higher than the opposite barrier because they may be adversely affected by the reflected sound.

3.5.5 Other Unique Design Considerations.

3.5.5.1 Overlapping Barriers.
Barriers which overlap each other (see Figure 20) are usually constructed to allow access gaps for maintenance, safety, and pedestrian purposes (see Section 9.4.1). A general rule-of-thumb is that the ratio between overlap distance and gap width should be at least 4:1 to ensure negligible degradation of barrier performance (see Figure 21). If a 4:1 ratio is not feasible, then consideration should be given to the application of absorptive material (see Section 3.4.1) on the barrier surfaces within the gap area.

Figure 20. Overlapping barriers (data base #5902).

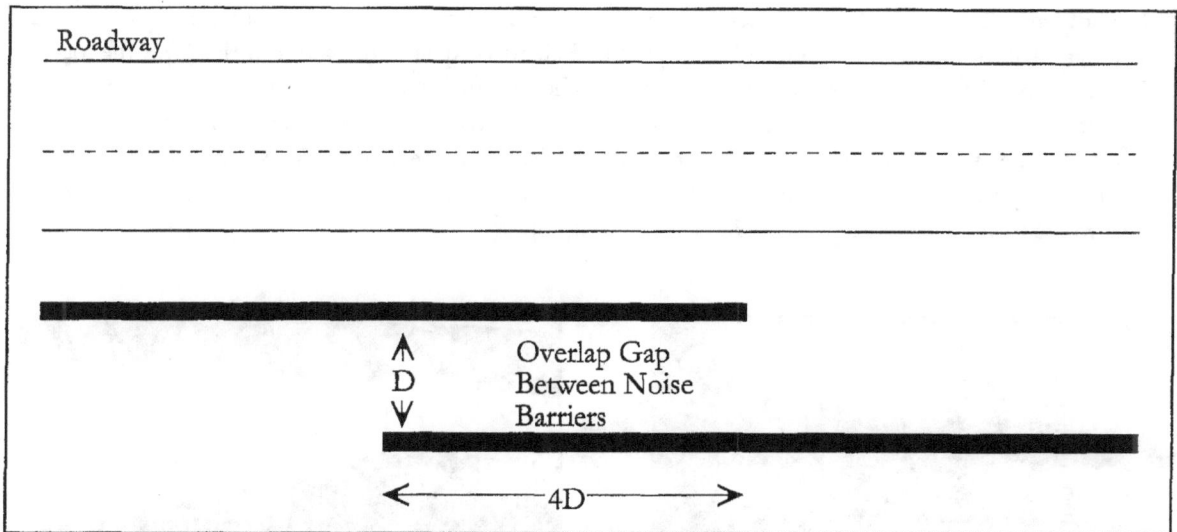

Figure 21. Overlapping barriers.

3.5.5.2 "Zig-zag" Barriers.
A barrier using concrete panels arranged in a "zig-zag-like" or "trapezoidal" configuration (see Section 4.1.2.3.1) is advantageous because it is structurally sound without the use of a **foundation**. This type of barrier can also be visually pleasing to motorists because it provides variation in form (see Figure 22). It does not, however, have any substantial additional sound attenuation benefits.

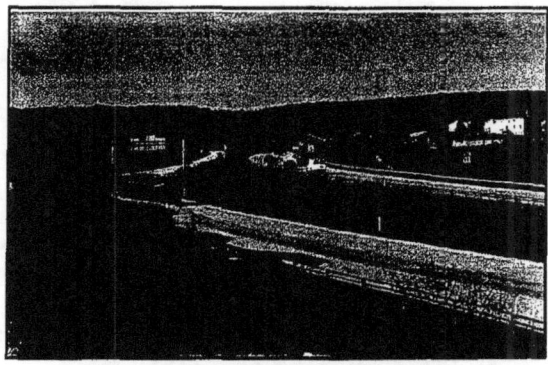

Figure 22. "Zig-zag" barriers (data base #8057).

3.5.5.3 Tops of Barriers.
There has been limited research into varying the shape of the top of a barrier (see Figure 23 and 24) for the purpose of shortening barrier heights and possibly attaining the attenuation characteristic of a taller barrier. The technical rationale is that additional attenuation can be attained by increasing the number of diffractions occurring at the top of the barrier. Shorter barrier heights could improve the aesthetic impact on communities and motorists by preserving more of the view.[18,38]

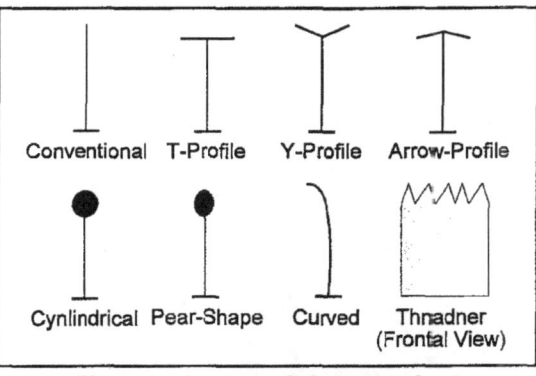

Figure 23. Special acoustical considerations: tops of barriers.

Studies have shown that a T-profile top barrier (see Figure 25) provides insertion losses comparable to a conventional top barrier when the difference in their heights is equal to the width of the T-profile top. When the two barriers are the same height, the T-profile top barrier has been shown to provide an additional 2.5 dB(A) insertion loss over the conventional top barrier. Y- and arrow-profile tops also performed better than conventional tops, however, to a lesser degree than the T- profile tops.[39,40] Cylindrical, pear-shape, curved, and Thnadner top barriers have not shown substantial benefits, unless an absorptive treatment was incorporated into the barrier tops.[41,42]

Figure 24. Special top of barrier (database #2395).

Figure 25. T-profile top barrier (database #1312).

Although there are some acoustical and aesthetic benefits associated with special barrier tops, the cost of constructing these shapes typically outweigh the cost of simply increasing the barrier's height to accomplish the same acoustic benefit.[43]

FHWA Highway Noise Barrier Design Handbook — Acoustical Considerations

Section Summary

Item #	Main Topic	Sub-Topic	Consideration	See Also Section	✓
3-1	Atmospheric Effects	Atmospheric Absorption, Refraction, Turbulence	Field measurements should not be performed when wind speeds are greater than 5 m/s, or when strong winds with small vector components exist in the direction of propagation.	3.3.1 14.1.2.1 15.1.2	
3-2	Barrier Design Goals	Barrier Sound Transmission	Barrier panel materials should weigh 20 kg/m² or more for a transmission loss of at least 20 dB(A).	3.4.2	
		Barrier Length	Ensure barrier height and length are such that only a small portion of sound diffracts around the edges	3.5.2	
		Wall vs. Berm	A berm requires more surface area, but provides 1 to 3 dB(A) additional attenuation versus a wall.	3.5.3	
		Reflective vs. Absorptive	Communities may perceive sound level increases due to reflections. Sound reflected between parallel barriers may cause degradations in each barrier's performance from 2 to as much as 6 dB(A), but in most practical situations, the degradation is smaller.	3.5.4	
		Overlapping Barriers	Ensure the ratio between overlap distance and gap width (between barriers) is at least 4:1.	3.5.5.1	
		Special Tops for Barriers	The cost of constructing these special shapes typically outweigh the cost of simply increasing the barrier's height to accomplish the same acoustic benefit.	3.5.5.3	

4. NOISE BARRIER TYPES

This section describes the differences between the following two basic types of noise barrier systems, as well as special features associated with each:

Ground-Mounted
and
Structure*-Mounted*

4.1 Ground-Mounted

Ground-mounted noise barrier systems are barriers constructed into or placed on top of the ground. This section will discuss the features of the three basic types of ground-mounted noise barrier systems:

- Noise berms (Section 4.1.1);
- Noise walls (Section 4.1.2); and
- Combination noise berm and noise wall systems (Section 4.1.3).

4.1.1 Noise Berms. Noise barriers constructed from natural earthen materials such as soil, stone, rock, rubble, etc. in a natural, unsupported condition are termed, noise berms (see Figures 26 and 27). These types of barriers are typically constructed with surplus materials available on the project site or from materials transported from an off-site location. The source and availability of such material are factors which can significantly affect the cost of such systems. Noise berms generally occupy more space than a wall type of barrier. This is mainly due to the sloping sides of the berms which must be gradual enough to maintain stability of the structure. For most berms, side slopes of 2:1 "run:rise" (i.e., 2 m horizontal to 1 m vertical) are typical, although on occasion steeper slopes (1½:1) may be acceptable. For berms constructed from rock (in an unsupported condition) side slopes as steep as 1:1 may be acceptable. The top of the berm may be of minimal width (with normal slope rounding) or it can be designed with a relatively wide plateau. While the level plateau area results in more space required to construct the berm, it provides for easier maintenance of the berm and offers an area for placement of such features as plantings, a **right-of-way** fence, or even a noise wall which could be used for improving the acoustical effectiveness by effectively increasing the height of the barrier system.

Figure 26. Noise berm (data base #2584).

Figure 27. Noise berm (data base #2712).

Preceding Page Blank

Other factors to consider in selecting a berm as a noise barrier system include:

- **Right-of-way requirements** - Is the existing right-of-way width sufficient, or is additional property required for its construction?

- **Location of the berm in relation to the right-of-way line** - Should the berm be constructed entirely on the roadway right-of-way, the adjacent owners' property, or with the right-of-way line running down the center of the berm?

- **Visual Implications both residential and highway side** - Is the berm going to be too ominous and overshadowing in a small residential back yard?

- **Destruction of existing features for construction of the berm** - Will it be necessary to remove older trees or other aesthetic features?

- **Maintenance and accessibility requirements** - Will the berm be left in a natural state or will it be mowed/landscaped and, if so, by whom?

- **Drainage implications** - Are special features necessary to avoid disrupting natural drainage patterns and possibly flooding adjacent lands or the roadway?

4.1.2 Noise Walls. Most noise wall systems are fabricated off-site; i.e., all of the components for this type of noise wall (foundation components excluded) are fabricated in a plant, then transported to the project site, and assembled on-site. Noise wall systems fabricated on-site include only cast-in-place concrete walls. This section includes a discussion of the following types of noise wall systems.

- Post-and-panel noise walls (Section 4.1.2.1);
- Brick and masonry block noise walls (Section 4.1.2.2);
- Free standing noise walls (such as precast concrete, planted/bin type, and stone crib) (Section 4.1.2.3);
- Direct burial panels (Section 4.1.2.4);
- Noise walls used to partially retain earth (Section 4.1.2.5); and
- Cast-in-place concrete noise walls (Section 4.1.2.6).

4.1.2.1 Post and Panel. This type of system consists of noise barrier panels mounted between foundation-supported posts (see Figure 28). Primary elements of this type of barrier include: post and post/foundation attachments, panels, and panel-to-post connections.

Figure 28. Post and panel noise wall (data base #27).

- **Post and Post/Foundation Attachments** - The type of post and post-to-foundation **anchorage** system is typically determined by the **responsible organization's** structural criteria for the particular type of barrier selected and the requirements of the system itself. The following describes some of the most common types based on currently constructed noise barrier systems:

  - *Reinforced concrete foundation with post bolted to top of foundation* - This foundation may be either concrete cylinder (**caisson**), **spread footing**, or **continuous footing** with either steel, reinforced concrete, or wood posts (see Figures 29 and 30).

Figure 29. Post and panel attachments: concrete cylinder (data base #528).

Figure 30. Post and panel attachments: continuous footing (data base #529).

  - *Reinforced concrete foundation with partial-depth embedment of post* - This foundation is limited to a concrete cylinder with posts being either steel or reinforced concrete. The posts are usually embedded in the reinforced concrete foundation to a point where the overlap of the post and the reinforcing is sufficient to achieve the desired structural strength in the foundation connection.

  - *Unreinforced concrete foundation with full-depth embedment of post* - This foundation is limited to a concrete cylinder with posts being either steel or reinforced concrete. The posts are usually embedded into an unreinforced concrete foundation to within 300 mm (1 ft) from the bottom of the footing (see Figure 31).

  - *Wood posts embedded in augured cylindrical hole with stone* **backfill**. - This type of foundation is typical of wood noise barriers systems.

Figure 31. Post and panel attachments: embedded post (data base #5).

- **Panels** - Noise barrier panels (or their components) are normally prefabricated and shipped to the project site. Size and configuration of panels vary depending on project application. With the exception of wood post systems and certain proprietary noise barrier products, the type of panel used is not generally dictated by the post type. For instance, steel posts are used to support wood, steel, and concrete panels; concrete posts can support concrete, wood, or steel panels. However, wood posts, commonly used for wood panel installations are seldom used for metal panels and never for concrete panels. Panels can be generally grouped into two categories - full height panels (including panels pre-assembled into full-height size) and panels stacked on site (see Figures 32 and 33).

Figure 32. Full height panel (data base #2946).

Figure 33. Stacked panel (data base #2680).

When considering the type of panel (full height or stacked), the following factors should be considered:

- *Post spacing* - Larger post spacings may be more economical than shorter spacings but may dictate the use of stacked panels due to shipping constraints. Longer, narrower panels are, typically, more susceptible to warping (in all directions) than shorter panels for some types of materials. The effect of such warping must be considered in light of stacked panel joints, post-to-panel connections, and visual implications (such as shadows caused by uneven joints).

- *Shipping requirements* - Due to highway vertical clearance restrictions, large full-height panels shipped in a vertical or near vertical position may have to be laid on their sides on trucks. This may dictate the placement of lifting inserts on both the top and sides of panels. If these panels are shipped in a horizontal position (flat on the truck), oversized load permits may be required with restrictions of horizontal clearances of highway underpasses and overpasses (see Figure 34).

Figure 34. Post and panel: shipping requirements (data base #527).

- *Weight and Size Limitations* - Some manufacturing facilities may be limited in terms of their capacity to manufacture, handle, and store larger panels above a certain size or width.

- *Acoustical considerations* - When designing noise barriers using stacked panels, careful consideration must be given to the design of the horizontal joints between panels. Connections must be such to preclude sound leaks due to gaps (see also Section 3.4.2).

- *Aesthetics* - Stacked panels by their nature create horizontal joints. The relationship of these joints to the aesthetic pattern of the barrier system must be considered (see Figure 35). A well-designed stacked panel system can integrate the panel joints into the aesthetic pattern of the barrier. Use of stacked panels with exposed aggregate surface treatment may emphasize the inherent variations between one panel and another. Use of stacked panels with a vertical form liner finish may require a deep pattern to help "hide" the horizontal joints. Care must also be taken to ensure

Figure 35. Post and panel: panel aesthetics (data base #2979).

proper alignment of one panel on top of the other when vertically oriented designs and textures are used such as **fluting**. This alignment may become compromised with time due to differential settlement of the ground. Stacked panel usage may limit the choices of aesthetic treatments on one or both sides of a concrete barrier. However, the joints may enhance the appearance if they are incorporated into the overall aesthetic design of the noise barrier wall.

- *Installation implications* - For any given post spacing, heavier capacity equipment (crane, lift, etc.) is usually required for a full height panel than for a stacked panel system. In certain restricted areas (such as, when a barrier must be placed under a **bridge** where access may only be available for smaller pieces of equipment, or where lifting height is restricted by overhead power lines) the use of stacked panels may be the only option (see Figure 36). On the other hand, the use of a full height panel requires only one lifting and setting operation for each **bay** of the barrier system and negates the need to align, seal, or caulk any horizontal joints between panels.

Figure 36. Post and panel: installation implications (data base #6049).

- *Maintenance considerations* - The possibility of a barrier panel being damaged and needing replacement should be considered when choosing a panel type. A damaging collision on a full height panel would require the replacement of the entire panel. A similar collision on a stacked-height panel bay may only require the replacement of one or a few damaged panels (see Figure

37). If the damage occurs at the bottom of the panel bay, it is likely that all of the stacked panels would need to be removed and then reset. Matching the new replacement panel (texture, color, etc.) with the remaining existing panels is a factor requiring consideration with either the full-height or stacked panel system, although the possibility for variance is multiplied with the use of a stacked panel system. In addition, the possibility of damaging intact panels is increased during the removal and reinstallation operation. Since replacement of barrier elements may be required

Figure 37. Post and panel: maintenance considerations (data base #5572).

throughout the life of the noise barrier, the availability of replacement parts becomes a critical issue. To address this concern, some agencies have instituted a stock piling policy where the contractor/manufacturer, at the time of construction, supplies additional components to the organization responsible for maintenance. This concern is discussed further in Section 12.2.

- **Panel-to-Post Connections** - A variety of techniques are available and have been utilized to attach or secure panels to posts. In selecting these connections, there are important acoustical and structural criteria that need to be considered. It is also essential that any factors related to the specific materials used in both the post and panel construction be evaluated in the design of the connection. This is especially critical in the design of barrier systems using dissimilar materials such as steel and aluminum and specialized materials, such as the connection between steel posts and transparent plastic or glass panels. Panel-to-post connections must be designed to eliminate any significant sound transmission leaks. As such, they should have a snug fit along the entire post-to-panel contact area or sufficient panel embedment into the post. These acoustical requirements may be accomplished by a variety of techniques related to the physical "fitting" of post-to-panel connection points, such as the use of **backer rods** and/or caulking and use of sound absorption material on the panel at the connection point. Any packing material used to ensure a snug fit at the connection point between the post and panel must be designed to remain in place and be functional for the life of the barrier. The connection should also be designed to eliminate or minimize any damage that could be caused through normal movement or vibration to either the post or the panel.

Important structural considerations related to the post-to-panel connections involve the following:

- *Panel-to-post **dead load transfer*** - For a typical post and panel system, this load transfer occurs as **point loads** at the base of each post (if the panel is supported by the bottom of the post) or at the top of each caisson (if the panel is supported directly by each caisson). For stacked panel systems, all dead loads should be assumed to be transferred to the bottom panel. Further, the bottom panel should be assumed to have no support underneath its entire length except for the two endpoints. Similarly, no support (except at endpoints) is assumed to exist along the bottom of full height panels in a post and panel caisson-supported system. However, if the post and panel system is supported by a continuous footing, then dead loads are assumed to be transferred uniformly through the panel (or bottom panel in a stacked panel system) along the entire length of the footing. Loads are transferred differently with systems using nailed, screwed, or bolted panel-to-post connections.

- *Panel-to-post wind load transfer* - The important consideration here is that the actual constructed panel-to-post connection is consistent with the design assumptions related to wind load transfer. If the wind load design assumed that loads were uniformly distributed from the panel to its two supporting posts along the full height of each post, then sufficient physical contact must be provided in the constructed system to ensure even load distribution from the panel to the post. If panels are wedged or blocked to the post at several locations, a full length load transfer may be ensured for loads applied in only one direction. However, the point loading resulting at the wedge points must be considered for loads applied in the opposite direction against the wedges or blocks. Full length load transfer may be provided by use of appropriate grouting, caulking, or mechanical devices which will ensure even distribution of loads. In the case of free floating panels between posts, the gap must be kept to a minimum and only sufficient enough to allow the panels to be installed without damage to the post or the panels and to ensure a continuous load transfer from the panel to the post.

4.1.2.1.1 Tilted Post and Panel. This type of system is a specialized type of post and panel system (see Figures 38 and 39) used in areas where sound reflections (either single or multiple) could be a problem if the more standard vertical noise walls were used (see Section 3.5.4). This type of system is most commonly manufactured using precast concrete elements, but at least one known system uses wood.

Figure 38. Tilted post and panel noise wall: community side (data base #36).

Figure 39. Tilted post and panel noise wall: highway side (data base #34).

Aside from the specialized loading treatments related to the tilted design, the considerations for post and panels discussed above apply to tilted post and panel systems. Since the angle of tilt is generally in the range of ten degrees, the issue of the wall being flat enough for people climbing it does not appear to be substantial. Because of aesthetics, care should be taken where such tilted walls transition to vertical walls or end abruptly. Similarly, the proximity of such tilted walls to residences and other areas of public use is probably more significant from a visual standpoint due to the possible perception of the wall "falling down."

4.1.2.2 Brick and Masonry Block. This type of system includes barriers constructed of fabricated brick or masonry block units (see Figures 40-43). Typically, these types of systems are constructed by laying the brick or masonry block in a conventional fashion using a continuous spread footing as a base. However, in certain instances, such barriers may be constructed on a base beam supported at the ends by the posts or by the top of the concrete caissons for the posts.

Figure 40. Brick noise wall (data base #8014).

Figure 41. Brick noise wall (data base #560).

Figure 42. Masonry block noise wall (data base #2454).

Figure 43. Masonry block noise wall (data base #2457).

4.1.2.3 Free Standing Noise Walls. This type of barrier system includes barriers which support themselves. Such barriers, constructed to date, can be grouped into the following general categories:

- Precast concrete (Section 4.1.2.3.1);
- "Planted" or Bin Type (Section 4.1.2.3.2); and
- Stone crib (Section 4.1.2.3.3).

4.1.2.3.1 Precast Concrete. These systems generally obtain their stability from the combination of their "zig-zag-like" or "trapezoidal" configuration and their system mass (see Figures 44 and 45). Depending upon soil conditions, precast free-standing walls may be supported by compacted soil and a well-drained stone base, a plain cement concrete leveling pad, or a continuous reinforced concrete footing.

Figure 44. Precast free standing concrete noise wall (data base #1206).

Figure 45. Precast free standing concrete noise wall (data base #1218).

4.1.2.3.2 "Planted" or Bin Type Barriers. These systems obtain their stability from a type of structural shell, typically either concrete, wood, or plastic, which is filled with soil and then planted (see Figures 46 and 47). These systems are most often supported by some form of continuous concrete leveling pad or footing. However, depending on the design and the type of plantings, these systems may be set directly on top of the existing ground with little or no preparation other than minor leveling.

Careful consideration needs to be given to the type of planting selected and to the means for providing adequate watering of the plant material during all seasons. Maintenance requirements can be significant on such systems, particularly related to items such as weeding, removing large saplings which grow from blown-in weed seeds (if not removed they can adversely affect the structural integrity of the barrier), and replacing pockets of washed-out soil. Safety, security and liability issues such as the ability to climb the steps of the planted wall should also be considered.

Figure 46. Bin type noise wall: plastic (data base #247).

Figure 47. "Planted" noise wall: concrete (data base #2567).

4.1.2.3.3 Stone Crib. This special type of barrier system (also referred to as a **gabion** system) is comprised of crushed rock contained in large rectangular baskets made of heavy wire mesh (see Figure 48). For aesthetic purposes, these wires can be coated with vinyl which are available in various colors. The baskets are stacked on top of each other in a pyramid fashion to obtain the required barrier height and stability. The baskets are typically placed on well draining, compacted ground. Their structure is flexible enough to tolerate some settlement. This type of system is only feasible if sufficient quantities of suitable rock material are readily available close to or on the project. Little, if any, plant life can be expected to grow on or within this barrier system. The system adapts well to rolling topography.

Figure 48. Stone crib noise wall (data base #549).

4.1.2.4 Direct Burial Panels. The direct burial panel type is a special panel system which involves burying a portion of one end of the panel (either precast concrete or wood) directly into the ground with no other means of foundation support (see Figure 49). With this type of system, the panels are usually full height and the connection to adjacent panels are typically designed as a tongue and groove system. Since differential settlement of the panels will most likely occur, a smooth top-of-wall profile cannot be expected. Therefore, a jagged profile should be considered during the design of this system (see Figure 50). In some cases, this differential settlement may not be even throughout the length of each panel, thus causing tilting of the panels and ultimately resulting in separating and gapping at the vertical tongue and groove joints.

Figure 49. Direct burial panels (data base #348).

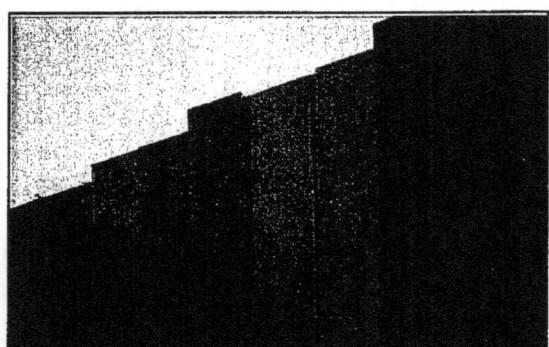

Figure 50. Direct burial panels (data base #351).

4.1.2.5 Noise Walls Used to Partially Retain Earth.
In certain instances it may be necessary and advantageous to utilize the bottom portion of a noise wall system to retain earth from either the residential or roadway side. Such applications have been successfully employed where barriers are constructed near the **slope hinge point** of a highway on fill and near the top of a highway cut section (see Figures 51).

Figure 51. Noise walls used to partially retain earth (data base #8027).

In such applications which have employed a post and panel system, the normal depth of the foundation for the noise barrier wall is usually sufficient to retain fills up to 500 mm (1.5 feet). Otherwise, careful structural analysis of the system is required to ensure that the bottom portion of the panel (if full height) or the bottom panels in a stacked panel system are sufficiently strong to retain the soil (see Figure 52). Panel-to-post connections and the joints between stacked panels require similar analysis. In these systems, special drainage designs may be needed to ensure that water does not "pond" or saturate the soil retained by the noise wall. In addition, the design should allow for proper drainage of the soil via weeping holes in the walls or via other suitable means.

Other types of noise walls (concrete block, free standing, cast-in-place, stone crib, etc.) may be considered for such earth-retaining applications, but only after the same careful consideration of the above factors. Combination **retaining wall**/noise barrier systems (generally requiring more significant depths of soil) are discussed in Section 4.2.2.1.

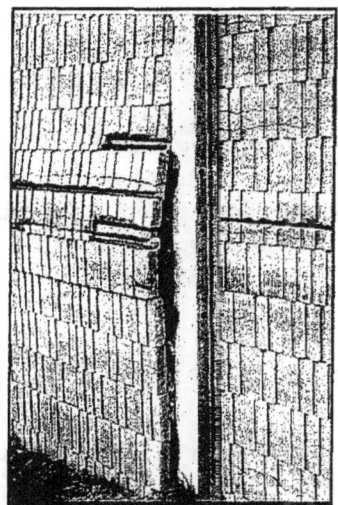

Figure 52. Noise walls used to partially retain earth (data base #480).

Potential also exists for the adjacent land owner to re-**grade** their property in the vicinity of the noise barrier. If the surrounding topography has the potential for regrading by adjacent owners, then it should be assumed that this will occur and the impact of this action must be considered in the design of the noise barrier.

4.1.2.6 Cast-In-Place Concrete Noise Walls.

These type of barriers are constructed at the project site (see Figure 53). The construction process includes excavating for the footing, erecting form work, setting reinforcement steel, pouring concrete, surface finishing, and curing. Except for certain foundation elements, such systems are significantly different from the noise wall systems discussed in the previous sections in terms of construction techniques, architectural and aesthetic treatment processes, and inspection and quality control procedures. All material testing and inspection procedures must be done in the field both during construction and on an as-erected product. All casting and curing occurs under a variety of weather conditions as compared to more controlled conditions for many of the components of prefabricated barrier systems.

Figure 53. Cast-in-place concrete noise wall (data base #1054).

Form liners and architectural inserts must be placed on vertical or near vertical surfaces of the form work which may lead to a significant increase in imperfections in the wall surface when compared to precast components usually cast in a horizontal position. In addition, the application of concrete-retarding chemicals to the vertical form work surfaces for the purposes of obtaining an exposed aggregate finish is significantly more difficult than in precast operations. Therefore, obtaining a consistent and acceptable exposed aggregate surface may not be possible. Surface textures obtained through raking, brushing, or stamping of concrete are not possible with cast-in-place walls.

4.1.3 Combination Noise Berm and Noise Wall Systems.

Many noise barrier systems consist of a portion of the barrier height obtained through use of an earth berm with the remainder of the required height achieved by placing a noise wall on top of the berm (see Figures 54 and 55). The foundation, post, and panel considerations for these wall portions are similar to those discussed above.

Figure 54. Combination noise berm and noise wall system (data base #27).

Figure 55. Combination noise berm and noise wall system (data base #993).

Additional considerations with this combination type of system relate to the following factors:

- **Reactive loadings on the berm's soil due to the wall portion** - The soil design parameters for earth in berms are substantially different than those of undisturbed soil. Factors such as the frictional interaction of the foundation and compressive loads on the berm's soils need to be considered in detail. For example: the area between the wall's foundation and the berm's side slopes need to be thoroughly analyzed in relation to the **shear slip circle**.

- **Plateau area on top of the berm** - An adequate level area on top of the berm should be provided to ensure stability of the soil at the base of the noise wall structure. Generally, a minimum plateau width of 2 m (6.5 ft) should be provided for foundation stability. This minimum requirement will help alleviate erosion of the berm resulting from rain hitting the wall and flowing down onto the berm.

- **Need for and location of right-of-way fence** - It is possible that the placing of a noise wall on top of a berm may negate the need for a right-of-way fence. However, it is also possible that a right-of-way fence may still be required at the right-of-way line or another location. This then would require consideration of maintenance and accessibility issues related to the area between the right-of-way fence and the noise wall commonly known as "no man's land" or "dead man's zone."

4.2 Structure-Mounted Noise Walls

This section discusses the types of noise walls used on structures, the concerns related to structure-mounted noise walls, and the general design and construction techniques used to address these concerns. There are two primary types of structure-mounted noise walls:

- Noise walls on bridges (Section 4.2.1); and
- Noise walls on retaining walls (Section 4.2.2).

Figure 56. Noise wall on a bridge (data base #2374).

4.2.1 Noise Walls on Bridges.

4.2.1.1 Types of Noise Walls on Bridges.

A number of techniques have been successfully employed to attach noise walls to bridges (see Figures 56 to 59). While somewhat different procedures and operations exist for attaching noise walls to existing bridges as compared to attachments to new bridges, the resultant attachment types are similar enough to be discussed under the following general categories:

- **Post and Panel Noise Barriers**

    On top of parapet - Such attachments usually include high strength bolts anchored to or embedded into the top of the parapet. On new

Figure 57. Noise wall on a bridge (data base #5231).

construction, such bolts are often set in the parapet form work prior to the concrete pour. In existing parapets, bolts may be anchored by mechanical fastening or chemical bonding (epoxy grout) methods. Depending on the type of noise wall material, these high strength anchored bolts and nuts are used to secure either a continuous horizontal beam (or angle) or vertical posts to the parapet. Noise wall panels or other components are then secured to the beam or posts to create the in-place barrier. Obtaining a smooth or desired top of barrier profile with such a system may require each panel to be custom made if the top of parapet profile is not smooth and/or consistent. Any bottom of barrier jaggedness or gapping, can be concealed by **flashing**.

Figure 58. Noise wall on a bridge (data base #1717).

Inserted into parapet - This method should only be considered for new bridges. Although not as common an attachment technique, posts have been inserted into the parapet itself (either prior to casting of the parapet, or after parapet casting) via insertion into precast holes within the parapet wall itself.

Figure 59. Noise wall on a bridge (data base #5090).

On outside face of parapet - Although suitable for existing and new bridges, it is particularly suitable for retrofitting of existing bridges. A rather common practice is to mount noise barriers onto the outside face of the parapet (Note: special consideration should be given in the situation of two parallel bridges, where a sizable gap between the bridges might compromise barrier performance). The barrier posts are usually attached to the parapet by one of four methods:

- *Mechanical anchoring system* - This type of anchor system consists of a wedge shaped nut which is inserted into a drilled or cast hole in the concrete parapet wall. As the bolt is turned, the nut is forced to spread and is wedged in the hole providing a solid anchor for the bolt to be sufficiently tightened. This system is limited in it's use and should only be considered for use in concrete, and should not be used in situations where the anchor will be exposed to constant vibrations from traffic and wind loading. In addition, any drilling into the parapet walls may diminish its bearing capacity, particularly if reinforcing bars are severed during the drilling operation.

- *Chemical anchoring systems* - This system basically consists of a two-part epoxy mixture adhesive inserted into a drilled or cast hole in the concrete wall and then mixed by spinning the bolt inside the hole. This method is more suited for older structures and for areas where the anchors are routinely exposed to vibrations. However, the same concerns regarding the severing of the reinforcing bars during the drilling operation (see Mechanical anchoring system above) should be considered before using this method. When this product first came onto the market, concerns were expressed regarding it's durability and long-term performance. These concerns appear to have been addressed by the industry, and use of this type of anchoring method is not restricted to specific applications.

- *Bolt through system* - The bolt through system uses long bolts which are inserted into holes either cast or drilled completely through the parapet walls. This method addresses most of the concerns associated with the durability of both the mechanical and the chemical anchoring systems. However, it is more destructive to existing structures and may diminish the bearing capacity of portions of the wall.

- *Cast-in-place bolts* - Although a less commonly used method, this anchoring system is considered to be the most effective and least destructive of all methods. However, this method should only be considered for new structures or where key areas of the structure are being rehabilitated. There may also be some difficulty in maintaining bolt location tolerances due to movement of the forms during pouring.

Additional barrier anchorage may be provided via angle iron mounted to the top of the parapet. On barriers constructed as part of a new bridge construction, the bridge slab may be extended beyond the outside edge of the parapet, providing additional dead load support for the barrier.

- **"Post-less" Panels** - Such systems use either concealed posts or no posts with the panels, typically mounted in the following manners:

  On top of parapet - For concealed post systems, the post-to-parapet connections are similar to those discussed above for the post and panel systems. For "post-less" systems, the panels (typically constructed of relatively lightweight materials) are attached via bolts to two parallel angle iron pieces mounted to the parapet.

  On outside face of parapet - Such systems are mounted in manners similar to those for post-and-panel systems listed above except that the panels themselves are bolted to or through the parapet. With this type of system, additional detailed care should be taken in the design of the horizontal joints between panels to ensure a leak-free noise condition and to maintain the consistent alignment of adjacent panels.

- **Masonry Block Noise Barriers** - These barriers are "laid up" in a manner similar to ground-mounted masonry block barriers except that their anchorage is to the protective concrete bridge parapet wall, which usually has the same shape as the standard concrete traffic barrier walls; i.e., Jersey barriers. The anchoring is via reinforcing bars extending out of the top of the parapet wall. The noise barrier wall can be further strengthened by inserting reinforcing bars and concrete within the voids of the masonry blocks.

- **Cast-in-place integral with parapet wall** - On occasions it may be necessary and appropriate to construct noise barriers integral with the bridge parapet wall. This type of structure-mounted noise barrier wall is more suitable where short height barriers can provide the desired noise attenuation or in situations where it may be the only possible option due to restrictions in erecting any other types of barrier systems.

- **On parallel supporting structure adjacent to parapet** - This type of structure-mounted noise barrier wall is not as common as other methods mentioned previously. This mounting system is particularly suitable for older or weakened bridges, where the structure (parapet wall, deck, and/or superstructure) is incapable of supporting the loads of the desired noise barrier system. A parallel supporting beam or similar structure may be built immediately adjacent to the existing structure. This structure would support the full vertical dead load of the noise barrier wall and all or some of the torsion load, if the beam and/or the wall were attached to the adjacent existing structure.

4.2.1.2 Effect of Noise Walls on the Structural Characteristics of an Existing Bridge.

Placing a new noise wall on an existing bridge adds a significant amount of stress on a structure caused be the additional weight and rotational loading for which the existing structure may not have been originally designed. This may result in the need to add additional girders, beams, and diaphragms; strengthen the existing bridge deck; or modify the existing parapet. Additional solutions which should be considered are reducing the weight of the noise wall by using light weight material or, only if absolutely necessary, by reducing the height of the wall or, ultimately, eliminating the construction of the wall. The latter should only be considered under absolutely severe situations.

Besides the obvious additional costs of such structural modifications (above the noise wall cost), other issues related to modifying an existing bridge include:

- Maintenance;
- Protection of traffic (both on and beneath bridge);
- Accessibility to areas requiring modifications;
- Bridge vibrations due to existing traffic;
- Vibrations from construction operations; and
- Potential environmental mitigation measures (related to painting beams or working over waterways or wetland areas).

4.2.1.3 Effect of Noise Wall on the Structural Requirements of a New Bridge.

While additional costs are still incurred (compared to the same bridge without a noise barrier), the ability to design the noise wall as an integral part of the overall structure addresses most if not all of the loading and traffic-related concerns discussed above.

4.2.1.4 Potential for Damage to Noise Wall From Vehicular Impact or Airborne Debris.

The proximity of the noise wall to the traveled portion of the bridge usually makes the wall considerably more susceptible to damage (compared to most ground-mounted noise walls). Such damage (see Figures 60 and 61) may be caused by vehicle impact, airborne debris such as stones, vehicle parts, snow removal operations, or material from salt spreaders in areas subject to snow fall.

Figure 60. Noise wall damage from vehicular impact (data base #5331).

Figure 61. Noise wall damage from airborne debris (data base #1281).

4.2.1.5 Potential for Damage and Injury in the Event of the Noise Wall or Parts Thereof Falling From the Structure.
Factors considered in addressing this concern include:

- Type and proximity of land use adjacent to or beneath the noise wall;
- Location of the noise wall on the bridge;
- Noise wall-to-bridge attachment details;
- Weight, composition, and shatterability of the noise wall component parts; and
- Any mechanisms (either internal or external to the noise wall) designed to retain noise wall components.

Current barrier design practice appears to indicate that these factors are considered to various extent. The barrier designer may choose to perform an evaluation of these factors in order to address any potential for design or safety-related concerns. The scope and complexity of an evaluation of the consequences of a vehicular impact on a noise barrier could range from an informal qualitative discussion to a detailed quantitative consideration involving probability of occurrence and statistical evaluation of risks related to the potential for structural failure and/or damage to people and property.

4.2.1.6 Other Safety-Related Concerns.
The proximity of bridge-mounted noise walls to traffic has raised concerns related to issues such as vehicular sight distance, barrier shading which increases potential for highway icing, and adverse effects on highway lighting. These issues are discussed elsewhere in this document (see Sections 9.2 and 9.8).

4.2.1.7 Maintenance Considerations.
Snow drifting and storage implications, restrictions to bridge inspection teams using bucket trucks to inspect beams, and maintenance of the noise wall itself including graffiti removal, noise barrier and structure damage repair, repainting, etc., are some of the concerns which should be considered during all stages of design and construction. These concerns, except for bridge inspection, are common to both ground-mounted and structure-mounted systems.

These concerns are generally greater for bridge-mounted noise walls due to their proximity to the roadway and accessibility limitations.

4.2.2 Noise Walls on Retaining Walls.

Retaining walls are typically constructed to retain a highway fill section (where the adjacent ground is lower than the highway grade) or to retain the adjacent ground (where the highway is in cut in relation to the adjacent ground). In either case, the possibility exists that the installation of a noise barrier wall may be warranted either as a part of the retaining wall or in the immediate vicinity of the structure.

Although the required height of a noise barrier system could be accomplished by placing an earth berm behind the retaining wall, such a condition is relatively rare and would be evaluated in the context of loads on the retaining wall. Therefore, the discussion in this section focuses on the variety of techniques used to construct a noise barrier wall on or near a retaining wall (see Figures 62 and 63).

Figure 62. Noise wall on a retaining wall (data base #2947).

Figure 63. Noise wall on a retaining wall (data base #531).

4.2.2.1 Combination Cast-In-Place Retaining Wall and Noise Barrier Wall.

Where the retaining wall is cast-in-place (see Figures 64 to 66), the necessary noise barrier height may be attained in the following manners:

- By extending the height of the retaining wall as a continuation of the cast-in-place structure;

- By installing one of the noise wall systems, such as the post and panel system, the "postless" system, a brick wall, or a masonry wall on top of the retaining wall as discussed in Section 4.2.1.1; or

Figure 64. Combination cast-in-place retaining wall and noise wall (data base #1844).

- By anchoring a noise barrier system to the sides of the retaining wall, using similar methods, as described in Section 4.2.1.1, for anchoring noise barriers to bridges.

Whether the retaining wall is new or existing, the structure must be capable of accommodating the additional dead loads and torsion loads of the noise wall. In addition to structural considerations, other concerns with this type of a retaining wall/noise barrier system are discussed in the following sections as they relate to specific types of installations.

Figure 65. Combination cast-in-place retaining wall and noise wall (data base #5004).

Figure 66. Combination cast-in-place retaining wall and noise wall (data base #329).

4.2.2.2 Noise Wall behind Cast-In-Place Retaining Wall. Placement of a noise barrier wall behind a cast-in-place retaining wall (see Figures 67 and 68) requires careful consideration of the load transfers (both dead load and torsion loads) on both the earth behind the retaining wall and the retaining wall itself.

Figure 67. Noise wall behind cast-in-place retaining wall (data base #1745).

Figure 68. Noise wall behind cast-in-place retaining wall (data base #1691).

As compared to a noise wall mounted directly on top of a retaining wall, additional considerations related to the area between the noise wall and the face of the retaining wall with this type of a system include:

- Maintenance and landscaping;
- Drainage; and
- Safety and security issues related to access and fencing.

These issues are discussed in more detail in Sections 6.2, 7.1, and 9.4, respectively.

4.2.2.3 Noise Wall on or behind Retained Earth System Type Retaining Wall.

Many instances exist where highway fills or cut slopes are retained by proprietary systems which use metal straps, grids, or other techniques to strengthen, reinforce, and/or retain the earth mass behind the wall system. In this type of retaining wall system, the "retained" earth mass is the retaining wall system's supporting medium, as opposed to the structure itself. While these systems typically incorporate a concrete facing of one form or another, this facing is not the primary structural element of the retaining wall system, and thus, cannot be relied upon to support the loads of a noise barrier. Without an independent footing (usually a spread footing which may be tied to the cap of the retaining wall) a noise wall cannot be placed on top of, attached to the side of, or installed directly behind the face of this type of retaining wall system.

The following is a discussion on two typical methods of overcoming these restrictions. These methods should not be considered all inclusive and are not intended to restrict innovative approaches.

- **Offset noise wall** - Any noise barrier wall installed adjacent to this type of retaining wall system should not be allowed to be supported, even partially, by the retained soil mass. Therefore, the noise barrier wall should be set back some distance from the retaining wall face.

- **Independent Foundation** - Caisson-type footings can be erected in the "original" ground prior to the construction of the retained earth system. Due to the amount of structural loading caused by this type of footing, the caissons are typically designed to assume that no support will be provided by the retained soil mass. To address this structural limitation, post length and caisson size and depth can be increased to allow the posts to be extended up to the ultimate height (the top elevation of the retained earth system) and then the retained earth system "laid up" with its straps, grids, etc. diverted around the locations of the noise barrier foundations. With this technique, the noise walls can be placed somewhat closer to the face of the retaining wall, although careful analysis of structural loading should still be performed.

With either of these two techniques, the same issues related to maintenance, landscaping, drainage, safety, and access referred to in Sections 6.2, 7.1, and 9.4 should be considered.

4.2.2.4 Noise Barrier Walls in Combination with or behind Pre-Manufactured Retaining Wall.
A variety of retaining wall systems exist which are comprised of pre-manufactured materials such as precast concrete, metal, and plastic components which are assembled on site (see Figure 69). Some of these systems may also incorporate earth within their structural facing. In general, noise barrier walls cannot be supported directly on top of the face of these systems. Thus, their construction must consider the same factors as discussed above for barriers located behind retained earth systems.

Figure 69. Noise wall caisson foundations behind pre-manufactured retaining wall (data base #6534).

4.3 Special Features
Situations and conditions often warrant the incorporation of special features into the construction of noise walls. These features may be required because of engineering, acoustical, and/or aesthetic reasons.

4.3.1 Caps.
For the purpose of this discussion, caps are considered to be separate elements of the barrier system applied to either the top of noise walls or to the top of the noise wall posts. The "cap look," which is accomplished as an integral part of the fabrication/construction of the noise barrier wall panels, is discussed further in Section 6.1.3.

4.3.2 Emergency Access Openings.
Special modifications of noise wall systems are often required in areas where the access from one side of the wall to the other is required (or anticipated to be required) on either a regular basis (as per maintenance personnel and/or equipment), or on an irregular basis (as per emergency personnel and equipment). The locations and frequency of such access openings is, usually determined by the responsible organization (DOT, emergency response unit, fire departments, police departments, etc.) on a case-by-case basis. Emergency and pedestrian access openings and doors are usually required to be identified on both sides of the barrier by signs denoting mile marker, station, or some other distinctive identifying feature. Section 9.4 discusses in detail several techniques used to provide such access openings in noise barrier walls.

4.3.3 Drainage Openings in Noise Walls.
Normally, water falling in the vicinity of a noise wall is carried longitudinally along the barrier in a drainage **swale** to a catch basin or inlet from which it is piped either under the wall or into the main roadway drainage system. In certain instances, it is necessary to allow water collected in front of or behind a noise wall to pass to the other side of the barrier. Design of such openings in the noise wall must ensure that their size and frequency are such as to not degrade the acoustical effectiveness of the noise wall system. A more detailed discussion is provided in Section 7.1.

4.3.4 Attachments to Noise Walls. A noise wall may, in certain situations, be the only possible location or the most feasible location for mounting or attaching other roadway-related elements. Examples of such elements include call boxes, highway speed limit or advisory signs, highway lighting fixtures, etc. Incorporation of such features should occur integrally with the design of the noise wall, not as an add-on element. Safety considerations related to clearances (both horizontal and vertical), electrical connections, sight distances, etc. are essential. This subject is discussed in more detail in Section 7.2.

Section Summary

Item #	Main Topic	Sub-Topic	Consideration	See Also Section	✓
4-1	Noise Berm	Aesthetic	Consider the visual implications on both residential and highway side and the landscaping required.	4.1.1	
			Consider the destruction of existing features for construction of the berm.	4.1.1	
		Drainage and Utility	Provide for adequate drainage requirements.	4.1.1	
		Safety	Consider right-of-way requirements.	4.1.1	
		Maintenance	Consider accessibility to and around berm and landscaping requirements.	4.1.1	
4-2	Post and Panel Noise Wall	Acoustical	Ensure that there are no sound transmission leaks between stacked panels and panel-to-post connections.	4.1.2.1	
		Aesthetic	Coordinate texture treatments with stacked panels so that the joints are either concealed by the pattern or become a part of the pattern.	4.1.2.1	
			Consider special concerns related to tilted post and panel designs.	4.1.2.1.1	
		Drainage and Utility	Overhead wires and other utilities may preclude the ability to use full height precast panels.	4.1.2.1	
		Structural	Provide for the impact of panel-to-post wind and dead load transfers.	4.1.2.1	
			Provide specialized loading treatments related to tilted post and panel designs.	4.1.2.1.1	
4-3	Free Standing Noise Wall: Precast Concrete	Maintenance	Consider landscaping access requirements.	4.1.2.3.1	
		Installation	For precast: consider size limitations, shipping requirements, traffic implications, reusability of precast panels, quality assurance process.	4.1.2.3.1 5.1	
			For cast-in-place: consider on-site material testing and inspection procedures during construction and on an as-erected product, and weather concerns for on-site casting and curing.	4.1.2.6 5.1	
4-4	Free Standing Noise Wall: Planted or Bin-Type	Aesthetic	Consider the type of plantings.	4.1.2.3.2	
		Maintenance	Ensure landscaping upkeep.	4.1.2.3.2	
			Consider litter implications.	12.7	
		Safety	Consider implementing a deterrent for climbing on barrier.	4.1.2.3.2	
4-5	Free Standing Noise Wall: Stone Crib	Safety	Consider implementing a deterrent for climbing on barrier.	4.1.2.3.3	
4-6	Direct Burial Panels	Acoustical	Consider possible separating and gapping at the vertical tongue and groove joints.	4.1.2.4	
		Aesthetic	Consider a jagged top-of-wall profile rather than a smooth profile.	4.1.2.4	
		Structural	Consider possible differential settlement of panels.	4.1.2.4	
4-7	Noise Walls Used to Partially Retain Earth	Structural	Consider loading concerns on the retained soil.	4.1.2.5	

Item #	Main Topic	Sub-Topic	Consideration	See Also Section	✓
		Drainage and Utility	Ensure proper drainage so that water does not "pond" or saturate the soil retained by the noise wall.	4.1.2.5	
4-8	Cast-in-Place	Aesthetic	Form liners and architectural inserts must be placed on vertical surfaces of the form work which may increase imperfections in the wall surface.	4.1.2.6	
			Application of concrete-retarding chemicals to the vertical form work surfaces for the purposes of obtaining an exposed aggregate finish is difficult.	4.1.2.6	
		Installation	Consider on-site material testing and inspection during construction and on an as-erected product.	4.1.2.6	
			Consider weather concerns for on-site casting and curing.	4.1.2.6	
4-9	Combination Noise Wall and Noise Berm	Structural	Consider reactive loadings on the berm's soil due to the wall portion.	4.1.3	
			Consider a plateau area on top of the berm.	4.1.3	
		Safety	Consider the need for and location of right-of-way fence.	4.1.3	
4-10	Noise Walls on Bridges	Structural	Consider weight stress and rotational loading.	4.2.1	
			Consider bridge vibrations due to existing traffic.	4.2.1	
			Consider bridge vibrations from construction operations.	4.2.1	
			Consider the impact of any parapet attachments.	4.2.1	
			Consider the type of anchoring system.	4.2.1	
			Consider top of barrier profile if the top of parapet profile is not smooth and/or consistent.	4.2.1	
		Safety	Ensure protection of traffic (both on and beneath bridge).	4.2.1.4	
			Consider the potential for damage and injury in the event of the noise wall or parts thereof falling from the structure.	4.2.1.5	
			Ensure adequate vehicular sight distance.	4.2.16	
			Consider barrier shading resulting in highway icing, and adverse effects on highway lighting.	4.2.1.6	
		Maintenance	Consider accessibility.	4.2.1.7	
			Consider snow drifting and storage implications.	4.2.1.7	
			Consider possible restrictions to bridge inspection teams.	4.2.1.7	
4-11	Noise Walls on Retaining Walls	Structural	Consider any additional dead and torsion loads due to the noise wall.	4.2.2	
		Drainage and Utility	Provide for adequate drainage requirements.	4.2.2	

5. NOISE BARRIER MATERIALS AND SURFACE TREATMENTS

A variety of materials may be used for noise wall panels and posts. This section provides details of some of the more common materials: including a description of the material, its features, examples of typical use, special considerations, typical quality verification, and regional considerations, where applicable. In addition, because the selection of a particular surface treatment texture can depend on a number of factors including aesthetic requirements of both sides of the barrier, constructability issues, maintenance concerns, and particularly the type of barrier material, this section also discusses barrier surface treatments. For example, selection of a form liner finish on both sides of a barrier could negate the ability to use horizontally cast precast barrier elements and require the use of either vertically cast precast elements or cast-in-place barriers.

5.1 Concrete

Concrete is one of the world's most common and versatile construction materials (see Figures 70 and 71). It is a mixture produced by combining Portland cement, coarse and fine aggregates, and water, and may also include specific additives to modify curing rate, air entrainment, strength, fluidity, and porosity. For cast-in-place operations, concrete is normally delivered on-site premixed by concrete truck, but for small quantities, it can be mixed on-site. For precast products, the plants usually have their own batch plants capable of providing sufficient quantities to match production.

Figure 70. Concrete noise wall (data base #634).

Figure 71. Concrete noise wall (data base #1239).

- **Features** - Almost half of the noise walls constructed in North America to date are made of concrete. The proliferation of the use of concrete is not without reason. Concrete, if formulated, cast (precast or cast-in-place), and cured properly, is considered to be one of the most durable materials currently used for many highway products, including noise barriers. It is rugged and able to withstand severe temperatures, intense sunlight, moisture, ice, and salt. It is a versatile material capable of being shaped, molded, and textured to take on the appearance of anything from weathered wooden boards to rock face to stone blocks to virtually any sculpted mural topic imaginable. Its mass, even at a thickness of only 12 mm (0.5 in.), is well within any Sound Transmission Class requirement (see Section 3.4.2).

Concrete products lend themselves well to coloring or tinting by either incorporating pigments into the concrete mix before pouring or afterwards by applying a stain onto the surface of the cured products. For more details, see Section 5.9.2.1.

- **Typical Use** - The versatility of concrete also extends to the shape and the size with which the panels can be produced (e.g., precast stacked panels, cast-in-place and precast full height panels, and precast concrete block). In addition, concrete allows for a complete range of installation techniques including post and panel, post integral with the panel, free standing, direct buried, and on top-of-spread footings, continuous footings, traffic barriers, and retaining walls. Cast-in-place concrete walls have been typically used on bridges and retaining walls because of their flexibility of design, high structural strength, and resistance to vehicle impact damage.

5.1.1 Special Considerations.

- **Mix Design** - There are two basic types of concrete mix produced, dry cast and wet cast. Both are generally composed of aggregate, Portland cement, and water. The major difference between the two is the amount of water used.

    - *Wet Cast Concrete* - Mix contains enough moisture to allow proper chemical reaction between all the ingredients in the mix to form a sufficiently strong bond between each other. The concrete mix must be allowed to set in the mold before it can be removed, typically about 8 hours. Wet cast concrete flows better in a mold than the dry cast mix allowing the use of finely detailed form liners.

        This type of mix is suitable for both cast-in-place and plant production products.

    - *Dry Cast Concrete* - Mix contains significantly less water and only enough to allow the mixture to retain its shape after being compacted into a mold. This allows the product to be removed immediately from the mold after pouring. The product is then hydrated by the introduction of steam during the curing process. Since it is not as fluid as the wet cast concrete, it is not suitable for panels requiring fine surface details. This type of mix typically has superior strength capabilities over the wet cast mix. It is much more suitable for mass production processes under controlled plant conditions.

- **Size Limitations**

    - *Precast* - Precast panel sizes are typically confined in one direction to approximately 4.5 m (15 ft) due to limitations in shipping, with no limit on length other than by size and weight for handling. The minimum thickness is usually directly related to the amount of **concrete cover** required over the reinforcing bars or mesh, but is typically about 100 mm (4 in) plus an additional 25 mm (1 in) in total to allow for the reinforcing and any surface texturing.

    - *Cast-in-Place* - The minimum wall thickness is approximately 150 to 200 mm (6 to 8 in) due to the space requirements needed to place and vibrate the concrete around the reinforcing steel.

- **Precast vs Cast-in-Place** - Precast panels can be erected quickly if crane and truck access are readily available. Traffic hold-ups can be minimized with off-site panel fabrication and landscape damage can be avoided by the use of proper sized cranes which can span over the landscaping when setting the panels. On the other hand, the presence of a crane and truck haulage unit together, which are necessary during erection of the panels, can become a traffic problem, sometimes necessitating a lane closure.

- **Reuse of Precast Panels** - Precast concrete walls have the potential to be relocatable and have been used for temporary walls as well as permanent installations.

- **Off-Site Casting Yard vs Off-Site Casting Plant** - Casting yards are typically established by the manufacturer as a temporary manufacturing facility for a specific project, usually to reduce shipping costs when the permanent casting plant is too far from the site. Casting plants inherently have a higher level of quality control in the manufacturing, handling, and hauling than can be achieved at a casting yard. As a result, a greater number of imperfections will appear in products supplied from a temporary casting yard and will have to be dealt with on site through a more active quality assurance process.

- **Regional Differences** - Concrete products are suitable for any climate condition imaginable. However, the emphasis on specific characteristics of concrete may vary from region to region. For example, in the northern regions, freeze/thaw and salt scaling resistance is critical. Where as, in the southern or warmer climates, **expansion coefficients** and proper curing practices are more important. In coastal regions, the emphasis would be on density which helps to repel the penetration of salt laden moisture.

5.1.2 Verification of Quality.

For all testing, it is important to select samples which are a true representation of the finished products or of the material(s) being used in the casting of the noise barrier components.

The following are standard tests, normally conducted at the casting plant, on concrete to verify overall quality and to confirm desired properties of the products. The tests discussed within this section are described in detail in Section 10 which discusses product evaluation of all types of barrier materials. Although most of these tests are suitable for both wet and dry cast concrete, some are more suitable for one type as compared with the other. Suitability of the test is noted where applicable.

- **Slump Test** (Suitable for Wet Cast Concrete Only) - This test determines the stiffness and consistency of freshly mixed concrete and, in general, is a good indicator of the amount of water in the mix.

- **Air Content** (Suitable for Wet Cast Concrete Only) - This test determines the amount of air in cured concrete. It is primarily a good indicator of durability of concrete which may be frequently exposed to freezing and thawing conditions.

- **Compressive Strength** - This test determines the maximum compressive strength of cured concrete samples.

- **Air Void Analysis** (Suitable for dry-cast concrete products) - This test determines the shape and size of air voids in cured concrete samples.

- **Freeze-Thaw/Salt Scaling** - This test is a combination of two tests, which determines, to some degree, the cured concrete's resistance to salt scaling and also to frequent freezing and thawing cycles. It is a very good indicator of the quality of the curing process.

- **Density** - Determining the density of the concrete material provided information related to the degree of **compaction** the concrete mix was subjected to in the mold before curing. The denser the product, the better the quality of concrete, assuming that a suitable mix design was used and the product was cured properly.

- **Water Absorption** - This test determines the amount of water the sample can absorb over a given time period. Generally, the more water absorbed, the poorer the quality.

- **Minimum Cover Over Reinforcing** - Panels and posts should be checked to ensure that the minimum concrete cover over the reinforcing is maintained during the casting operation. Adequate cover is critical in preventing or slowing down the penetration rate of salt laden moisture from reaching the reinforcing bars. This results in the corrosion of the bars and subsequent **spalling** of the concrete surface along with the drastic deterioration of the structural properties of the components.

- **Dimensions** - All precast or cast-in-place concrete products should be checked for proper dimensioning of key features (see Section 11.5.1).

- **Visual Inspection** - All precast and cast-in-place concrete products should be visually examined to identify any unusual and unwanted features which will affect the structure, durability, and performance of the noise barrier wall, such as honey combing, knuckling, **cracks**, and voids.

- **Color Consistency** - A consistent color from panel to panel may be difficult to achieve; however, it is an important aesthetic factor in achieving a successful barrier system. To ensure a panel-to-panel color consistency, a surface-applied stain may be more effective than the use of integral colors or pigments in the concrete mix.

5.2 Brick and Masonry Block

Brick (see Figure 72) is typically manufactured using a clay and sand mix which is fired in a kiln to increase the brick's strength and durability. Bricks can be produced in varying sizes but most commonly in 70 x 95 x 200 mm (2 ⅔ x 3¾ x 8 in).

Masonry block (see Figure 73) is manufactured using a dry-cast concrete mix. These blocks can be produced in any size but with the most common in the range of 200 to 300 mm (8 to 12 in) thick by 200 to 250 mm (8 to 10 in) high and 355 to 460 mm (14 to 18 in) long.

- **Features** - Both brick and masonry block walls can be either hand-laid or preassembled by machine. Hand-laid walls have greater versatility in their ability to conform to the variety of ground contours encountered in the roadway environment and in their layout than do the preassembled panels with their fixed panel sizes and heavy equipment requirements. Preassembled panels have an advantage in speed of erection, provided that the site environment allows for easy maneuvering of the necessary cranes and transport vehicles. In addition, brick and masonry block walls can be constructed satisfactorily with no special leveling courses on grades of up to 6 percent. In some cases, brick is used as a facing or veneer on masonry block or cast-in-place walls.

Figure 72. Brick noise barrier (data base #6511).

Figure 73. Masonry block noise barrier (data base #2456).

5.2.1 Special Considerations. All brick and masonry walls, whether they are hand or machine laid, require a continuous concrete foundation (see Figure 74). The wall must be anchored to the foundation with reinforcing bars. Vertical and horizontal reinforcing bars are also needed in the wall itself to provide structural strength. Preassembled panels usually must be braced while the supporting concrete gains its strength.

- **Scaffolding for Installation** - In most cases, scaffolding is needed to install brick and masonry block noise barriers (see Figure 75). Cranes may be used to install prefabricated or preassembled panels, but crews are still needed on scaffolding to fasten the panels to the posts and framework. Scaffolding needs room, a good solid foundation, and a considerable amount of effort and time to install. All of these factors should be considered before this type of material is selected for a specific site.

Figure 74. Barrier concrete foundation (data base #568).

- **Regional Differences** - Masonry blocks are less suited for the more northern regions where the blocks would come into frequent contact with substantial amounts of deicing salts, which tend to deteriorate the blocks and the bonding mortar between them.

5.2.2 Verification of Quality. The tests discussed within this section are described in detail in Section 10 which discusses product evaluation of all types of barrier materials.

- **Compressive Strength** - The compressive strength of the brick or masonry block, the concrete used to fill the voids inside the wall, and the concrete used in the foundation should be tested, since these are the structural components of the wall system.

Figure 75. Scaffolding for barrier installation (data base #2455).

- **Dimensions** - Since concrete blocks and bricks are shipped on pallets, it may only be necessary to check the dimensions of one unit per pallet (see Section 11.5.1).

- **Mortar** - The mortar used in most concrete block noise barriers is an integral part of the structural strength of the wall enabling it to withstand lateral forces against the wall. Therefore, the quality of the mortar becomes quite critical and should be checked to ensure adherence to project specifications.

5.3 Metals

Three type of metals are most commonly used: (1) steel; (2) aluminum; and (3) stainless steel.

Steel - Steel is the least expensive and most common of all metals used in construction (see Figure 76). It is composed of a mixture, in varying proportions, of iron ore, carbon, and small amounts of other metals depending on the physical characteristics desired.

Most steel panels, posts, and **girts** are either: (1) coated with plastisols, bonded powders, enamel paints, or galvanizing material; or (2) manufactured as a self protecting **weathering steel**.

Figure 76. Metal noise barrier (data base #158).

Aluminum - Aluminum is a lightweight alloy commonly made from bauxite and is typically coated with a bonded powder, enamel paints or anodized (see Figure 77). It is not compatible with galvanized coatings.

Stainless Steel - Stainless steel is a highly durable and corrosion resistant metal alloy. It is a mixture of steel carbon, nickel, and chrome (in varying proportions). Since this material is virtually corrosion resistant, the surface does not need to be coated.

- **Features** - Metal panels have a weight advantage which makes them particularly useful for vertical extensions of existing sound walls, for mounting on existing retaining walls which have limited residual strength, or on bridge structures, because of their light weight.

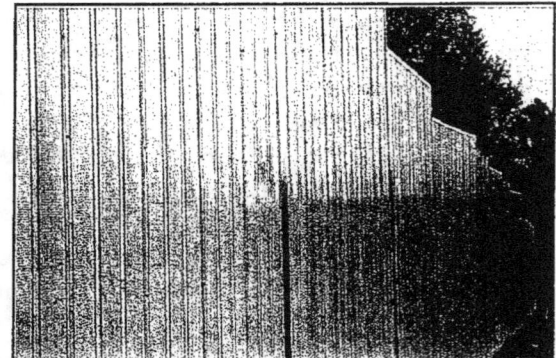

Figure 77. Metal noise barrier (data base #5692).

- **Typical Use** - This type of material can be used anywhere. However, bridges and retaining walls are ideal locations for the use of these light-weight type of panels.

5.3.1 Special Considerations.

- **Weathering Steel** - Such unpainted, rusting panels are found to stain adjacent concrete.

- **Adjacent Vegetation** - Plantings do not grow well next to sun-heated metal panels. In addition, vines have difficulty in gaining a foothold on coated panels.

- **Non-compatibility of Various Metal Combinations** - Care should be taken to ensure that differing metal which come in contact with each other do not have an adverse effect on one another. This is particularly true for aluminum coming in contact with steel. The aluminum acts similarly to the zinc in the galvanizing material where it is the sacrificial element and will eventually disintegrate over a short period of time.

- **Sound Transmission Class (STC)** - Most metal sheeting materials do not meet the typical minimum panel weight and/or sound transmission class required in typical noise barrier specifications (see Section 3.4.2). However, adding corrugations or ribs to the profile of the panel material tends to improve the sound transmission class of the panel.

- **Appearance** - Occasionally, metal walls impart an industrial appearance which is considered undesirable by some residential neighborhoods. This is particularly true for the back side where the girts and posts are exposed to the view of the residents. To overcome this perception, a double faced wall may be used. This system incorporates the use of steel panels on both sides of the post and framework.

- **Climbability** - The girts, in combination with horizontally mounted corrugated or ribbed metal panels, provide an opportunity for the wall to be climbed from the side containing the girts or ribs (see Figure 78). This can be prevented by the addition of a sheet of metal facing to cover the girts and posts on this side of the wall.

- **Glare** - All metal barriers are susceptible to glare from opposing light sources. This issue is addressed in more detail in Section 9.6.

Figure 78. Metal barrier: climbability (data base #157).

- **Conductivity** - Since all metals are electrically conductive, the installation of metal noise barrier walls should be avoided near power lines unless all metal components can be properly grounded.

- **Scaffolding for Installation** - In most cases, scaffolding is needed to install metal noise barriers. Cranes may be used to install prefabricated or preassembled panels, but crews are still needed on scaffolding to fasten the panels to the posts and framework. Scaffolding needs room, a good solid foundation, and a considerable amount of effort and time to install. All of these elements should be considered before this type of material is selected for a specific site.

- **Panel Thickness** - Metal noise barrier panels are typically in the range of 18 to 22 gauge thickness making them quite susceptible to damage from vandalism, debris, errant vehicles, snow plow operations, and other maintenance equipment. Therefore, consideration should be given to the thickness of the panel, the structural strength that can be achieved through corrugations, and the distance from the roadway that they are installed.

5.3.2 Verification of Quality.

The tests discussed within this section are described in detail in Section 10 which discusses product evaluation of all types of barrier materials.

- **Dimensions** - The panel, profiles, size, and gauge should be verified since any deviation from that specified in the design plans will effect the structural strength, durability, and performance of the noise barrier system (see Section 11.5.1).

- **Coating Thickness** - Coating thickness, whether it is a galvanized, painted, sprayed, or dipped, must be verified to ensure durability.

- **Accelerated Weathering** - This test provides information on how well coatings withstand extreme moisture, high temperatures and harsh light conditions. The test method normally involves exposing the samples to prolonged water spray and high doses of ultraviolet light, all carried out under high temperatures (approximately 63 degrees Celsius or 145 degrees Fahrenheit). Although accelerated test methods are not true representations of actual conditions, it does provide a reasonable tool to compare product performances and to be able to reasonably predict the results of long term exposure to harsh climatic conditions.

- **Corrosion Resistance** - This is another procedure used to test coatings, but is used mostly for testing their effectiveness in preventing corrosion of metal surfaces to which they are applied. The typical test method subjects a coated metal sample to constant exposure of salt and moisture.

- **Structural Strength** - Single, thin, flat sheets of metal usually do not have the structural strength to resist the wind loads to which noise barrier panels are normally subjected. Therefore, panels made of this type of material may require some form of stiffening before they are able to meet local structural design and testing procedure requirements.

- **Metal Properties** - Brittleness, hardness, and tensile strength should be verified by appropriate standard test method. Mill classification certificates should be available for all metal components.

5.4 Wood

Most wood noise barrier walls are constructed of **pressure preservative treated** lumber, plywoods (see Figure 79), and **glue laminated** products (see Figure 80).

Figure 79. Plywood noise barrier (data base #657).

Figure 80. Glue laminated post and plank noise barrier (data base #736).

A number of different species of wood have the potential for being used as a noise barrier product, but this does not mean that all perform equally. Some species, such as the pines, are well suited for pressure treatment. Whereas, it may be difficult to obtain a deep, uniform penetration of the preservative in spruces. Some of the more common species used are as follows:

Pacific Coast Douglas Fir	Eastern White Pine
Interior Douglas Fir	Lodgepole Pine
White Fir	Western White Pine
Western and Eastern Hemlock	Southern Yellow Pine
Western Larch	Red Spruce
Jack Pine	White Spruce
Red Pine	Poplar
Ponderosa Pine	Red Alder

- **Features** - Panels can be either installed piece by piece in the field or partially assembled in a plant or on the ground prior to attachment to the post. Power nailers, which are commonly used in the plant as well as in the field, make quick work of assembly. Some wood barriers can also be easily dismantled if future highway changes are needed. This material blends well with natural or residential background and does not conduct electricity.

5.4.1 Special Considerations.

- **Safety** - Consideration to safety issues, such as shatter resistance, should be given when mounting wood noise barrier walls on traffic barriers.

- **Burning Characteristics** - Wood noise barrier walls will burn under the same conditions as any other wooden fence. The smoke and emissions that are generated from burning treated wood are considered toxic. The ash left from the burning of this type of wood is also toxic and can leach into the surrounding soil and water supply.

- **Warping and Shrinkage** - Wood products are not dimensionally stable and tend to warp/shrink leaving open cracks between joints, particularly if they have not been properly seasoned or kiln dried. The thicker the wood products are, the more problematic warping can become.

- **Tongue and Groove Planking** (see Figure 81) - To prevent the occurrences of gaps between planks as the walls weather and the planks shrink or warp, specifications for wood plank walls should include deeper tongues and grooves than the industry standards.

- **Pressure Treating** - Most wood will **decay** rapidly when in contact with moisture. To combat this, the common practice is to pressure treat the wood with a chemical preservative. There are several acceptable chemical solutions used, all with relatively equal performance. Some of the more common ones are ACC, ACA, CCA, Penta, and Creosote.

Figure 81. Wood barrier: tongue and groove planking (data base #411).

  - *ACC* - Acid Copper Chromate.

  - *ACA* - (Ammoniacal Copper Arsenate) is a bright greenish colored waterborne preservative.

  - *CCA* - (Chromated Copper Arsenate) is a mild green colored waterborne preservative. This cannot be used on fir wood.

  - *Penta* - (Pentachlorophenol) comes in either a gas or oil borne formula. Colors vary, with the gas borne generally being lighter in color. This product has a distinct odor for a season and tends to draw out natural pitch leaving deposits and streaks on the wall facing.

  - *Creosote* - is a Coal Tar based oil borne preservative, usually dark brown in color.

- **Glue laminated posts and planks** - Due to the chemical composition of the glues commonly used to fabricate glue-laminated post and planks, these products are not suitable for treatment with water-based preservatives, only the oil-based Penta is recommended for this product.

- **Cutting** - Cutting pressure treated wood will expose untreated interior portions to the elements. These areas should be retreated with a compatible brush-on preserving solution.

- **Fasteners** - Fasteners used to assemble a wood noise barrier can either be staples, nails, lag bolts, carriage bolts, nuts and bolts, or screws. Ideally, they should be made of a non-corroding metal such as stainless steel or aluminum. However, aluminum and steel have been known to react unfavorably with some types of pressure-treating chemicals; steel nails are also susceptible to this type of chemical reaction.

- **Scaffolding for Installation** - In most cases, scaffolding is needed to install wood noise barriers. Cranes may be used to install prefabricated or preassembled panels, but crews are still needed on scaffolding to fasten the panels to the posts. Scaffolding needs room, a good solid foundation, and a considerable amount of effort and time to install. All of these issues should be considered before this type of material is selected for a specific site.

- **Wood Posts** - Large posts have a tendency to **check** or **split** open, exposing untreated surfaces to decay and attack by **insects**. Checking can be minimized by allowing the posts to season properly before being pressure-treated. In addition, the butt end of the post should be **kerfed** to allow deeper penetration of the preservative solution.

- **Color** - Initial color is a problem with timber walls since it is usually governed by the type of preservative chosen. These barriers will eventually fade to a weathered brown or gray color. The waterborne preservatives are initially green with some fading rapidly and more uniformly than others. Repairs to damaged sections will be conspicuous unless repair planks are acquired at the time of construction and allowed to weather in the yard.

Some saw mills stain their products with a preservative having an identifying hue or "mill bright." This practice creates visual chaos when different-sourced products are mixed in one installation.

5.4.2 Verification of Quality.
The tests discussed within this section are described in detail in Section 10 which discusses product evaluation of all types of barrier materials.

- **Structural Grade** - Specifying a good structural grade of lumber does not guarantee that all pieces of wood will be straight enough to permit the tight fit normally required for wood barriers. Therefore it is essential to visually confirm the grade of the wood used and removing any pieces that are warped, checked, split, or have excessive **knots**.

- **Dimensional Stability** - Structural-graded lumber does not ensure that the product will never shrink, particularly if the wood has not been properly seasoned or kiln dried before pressure treating. Therefore, all wood components, particularly large members, should be checked for dimensional tolerances (see Section 11.5.1).

- **Determination of Penetration** - This test determines the depth of penetration of the preservative into the wood. The penetration rate may vary between species.

- **Moisture Content** - This test determines the amount of moisture in the wood. It is a nondestructive test and should be conducted on all larger pieces and those where warping, checking, and splitting may be unpreventable or may have a serious impact on the overall performance of the wall.

5.5 Transparent Panels

The typical transparent noise barrier (see Figure 82) may use panel materials made of either glass or plastics such as acrylics, polymers, and polycarbons. Glass panels are commonly made of single-tempered or laminated-tempered glass sheets. Both plastics and glass can be tinted and can also be etched or given a frosty appearance.

Tempering of the glass is a heat-treating process which strengthens the glass, producing a much more shatter-resistant product. When it does shatter, the shards are small and granular in appearance, with pieces typically not larger than 12 mm (½ in). These are much safer

Figure 82. Transparent panel noise barrier (data base #1981).

than the long knife-like shards produced from shattering common non-heat treated glass. In addition to the tempering, the glass panels can also be laminated. This type of glass panel is produced by adhering two sheets of tempered glass sheets to both sides of a clear rubbery type flexible sheeting. When this type of glass panel is shattered, the glass will break into small granular-like pieces, where the pieces will remain adhered to the sheeting.

- **Features** - The transparent panel materials are an ideal way of reducing or virtually eliminating the visual impact of a noise barrier.

- **Typical Use** - Transparent barriers are normally only built for 3 reasons:
    - To prevent hindering the scenic view for the driving public;
    - To prevent hindering the scenic view for the residents adjacent to the roadway; or
    - To prevent hindering the view of retail establishments for the driving public.

Since transparent noise barriers costs can be as much as 20 times that of common concrete or steel panels, the decision to use transparent noise barriers should not be made lightly. Possibly, the only other reasons for their use would be to improve safety, where opaque noise barrier walls may have an adverse affect on stopping sight distance, visibility in merge areas, lighting, and shading.

5.5.1 Special Considerations.

Transparent noise barriers come with their own unique set of engineering, safety, and environmental considerations which are significantly different than most other types of material normally used for noise barrier panels.

- **Vandalism** - Plastic panels are particularly susceptible to vandalism (see Figure 83), not only from the typical paint can, but also from knives, lighters, or matches.

Figure 83. Transparent panel barrier: vandalism (data base #1947).

- **Mounting** - Depending on the type of material used for the panels, (glass, acrylic, etc.) and the size, the method of mounting can vary significantly. The general principle for mounting this type of thin, flat sheeting is that the method used must sufficiently reduce or remove the stresses on the material to eliminate the possibility of the panels breaking or falling out from between the supports. It is recommended that the manufacturer be contacted to obtain the best mounting method to be used for their specific product on every site.

- **Dimensions** - In order to enhance transparency, it would be preferable to use large pieces of material and to limit the number of supporting brackets. While transparent plastic sheeting materials are available in lengths exceeding 5 m (16 ft), the same is not true for tempered or laminated glass. Manufacturing constraints limit the maximum dimensions of glass plates. In addition, as the area of the panel increases, the thickness must also be increased to maintain structural integrity, thus increasing weight and cost (see Section 11.5.1). To reduce the need for thicker material, it is common to use smaller post spacing and/or framed panels that can be stacked between posts. Size of panels is also limited by handling capabilities.

- **Edge Conditioning** - To avoid thermal and stress cracking, all edges must be smooth and without defects. This is extraordinarily critical if the edges are not cut in a straight line.

- **Ultraviolet Light Stabilizers** - Some of the transparent plastic sheeting materials are sensitive to ultraviolet light. If exposed to sunlight for extended periods of time without being protected by UV stabilizer additives or coatings, the sheeting will haze and discolor, leaving them translucent or even opaque in some case. Even with the stabilizers, the sheeting will eventually be affected by the light. Acrylic-based sheetings are much less sensitive to sunlight and tend to stay transparent for a longer period.

  Note that glass, by itself, is not affected by sunlight. However, if laminating material is used, this material may be sensitive to ultraviolet light.

- **Shatter Resistance** - Although most commonly used plastic products are relatively shatter resistant, glass is not; even when the glass is tempered and/or laminated, the panel will shatter.

- **Glare** - All transparent barriers are susceptible to glare from opposing light sources. This issue is addressed in Section 9.6.

- **Road Debris Damage** - These types of panels are more susceptible to damage from flying debris than most other types of barrier materials. They are also very susceptible to the abrasive damage caused by the sand blasting action from stirred up road dirt.

- **Cost** - Depending on the size and type of material selected, the cost of transparent barrier sheeting can be anywhere from 10 to 20 times more than that of other barrier panel materials. However, in some areas of the country where the use of transparent barriers are prevalent, the costs may be much lower, even comparable to that of other barrier panel materials.

- **Cleaning** - To maintain their transparency, these types of panels need to be washed on a regular basis. This is of particular concern if the wall or the individual panels are tilted, which tends to hinder the natural cleaning process provided by rain on the underside of the panel. Access for cleaning of the panels is normally not a problem on the traffic side, which is usually the dirtier side of the wall. However, the opposite side may not be as accessible, and, in some cases, cleaning may not be feasible at all. This limitation should be considered when selecting barrier material. Cleanliness is particularly critical if the transparent noise barrier was constructed for safety reasons such as to improve visibility for stopping sight distance or merging.

- **Breakage** - Damaged panels cannot be repaired by patching. The only option is to replace the damaged sections.

- **Regional Differences** - There are generally no climatic restrictions for the use of any of the transparent sheeting materials.

5.5.2 Verification of Quality.
The tests discussed within this section are described in detail in Section 10 which discusses product evaluation of all types of barrier materials.

- **Dimensions** - The panel, profiles, size, and thicknesses should be verified since any deviation from that specified in the design plans will effect the structural strength, durability, and performance of the noise barrier system (see Section 11.5.1). Of particular importance with the transparent sheeting material, whether it is glass or one of the plastics, is the strict adherence to the tolerances for the mounting hardware and the sealants to avoid uneven stress points which can result in material fracturing or warping. Such consequences can occur if a panel is placed between two posts with insufficient room for expansions or inappropriate expansion or caulking material. If too much room is provided at the panel-to-post connection points, the panel can actually become separated from the posts under certain conditions which will result in substantial contraction of the barrier material and/or excessive panel movement caused by vibration or wind.

5.6 Plastics
There are several types of plastic materials available for use as a barrier material, including Polyethylene, PVC, and fiberglass (see Figure 84).

Figure 84. Plastic noise barrier (data base #782).

- **Features** - The most unique features of plastic products are their versatility and moldability. This material can be produced to perform and appear the same as almost any construction material on the market today. Its lightweight nature improves ease of handling both in the plant and in the field. In addition, most of these products are recyclable.

- **Typical Use** - Plastic noise barrier panels can be installed in almost any situation. However, due to their lightweight characteristics, they are particularly suitable for structure-mounted applications.

5.6.1 Special Considerations.

- **Burning Characteristics** - Plastic noise barrier walls tend to be more flammable than barriers made of other materials. The smoke and emissions that may be generated from burning plastics should be considered toxic. The ash left from any burnt material should also be considered as toxic and can leach into the surrounding soil and water supply.

- **Shrinkage** - Some plastic products are not dimensionally stable and may tend to shrink leaving open cracks between joints or may be susceptible to accelerated **creep** and deformation.

- **Ultraviolet Protection** - Some plastic products are very sensitive to ultraviolet light and tend to cause rapid deterioration of pigments, surface appearance, and material strength. To avoid this, it is possible to slow down the deterioration process by adding ultraviolet protection into the composition of the plastic at the time of molding.

- **Creep** - Creep, which is evident in most plastics to varying degrees, should be considered during the design of the barrier system by reducing the amount of strain to which the plastic components may be subjected.

- **Vandalism** - Plastic panels are particularly susceptible to vandalism from paint, knives, and lighters.

- **Shatter Resistance** - Although most commonly used plastic products are relatively shatter resistant, this characteristic tends to deteriorate over time, and the product becomes more brittle and may shatter on impact by flying objects or vehicles. Damaged panels can usually not be repaired by patching. The only option is to replace the damaged sections, thus increasing the cost of repairs and possibly jeopardizing the appearance of the barrier if similarly molded panels are no longer available or are difficult to reproduce at a reasonable cost.

- **Glare** - Depending on the surface texturing applied to the plastic surfaces, the barrier panels may be susceptible to glare from opposing light sources. This issue is addressed further in Section 9.6.

5.6.2 Verification of Quality.

There are countless standard test methods published to assist in the verification of plastic products. Each of these test methods are normally only relevant to very specific plastic formulations. The tests discussed within this section are described in detail in Section 10 which discusses product evaluation of all types of barrier materials. This handbook will not attempt to list or describe these and suggests current testing information be obtained from other sources.

- **Dimensions** - The panel's profile, size, and thickness should be verified since any deviation from design specifications will affect the structural strength, durability, and performance of the noise barrier system. Of particular importance is the strict adherence to the tight tolerances for the mounting hardware and the sealants to avoid uneven stress points (see Section 11.5.1).

5.7 Recycled Rubber

The issue of using recycled rubber from tires in products used for roadway construction has been under investigation for many years by numerous government agencies, world wide. The results of their efforts indicate widely varying success in trying to adapt this type of material into a usable product (see Figure 85).

Recycled rubber can refer to a wide range of products, made from an equally wide range of rubber compounds. In practice, however, the rubber waste stream is dominated by scrap tires. There are two other significant sources: (1) tire trim and off-spec tires from tire production, and (2) buffings from rubber product manufacturers.

Figure 85. Recycled rubber noise barrier (data base #3124).

5.7.1 Special Considerations.

- **Flammability and Smoke** - Rubber is notorious for its high flammability and the dense smoke which is produced when it burns. If a noise barrier made from this material should ignite as a result of such incidents as grass fires, accidents, or vandalism, the accelerated flamespread and the dense smoke produced could result in safety and legal issues. To reduce rubber's susceptibility to these concerns, flame and smoke retardants are available that can be added to the mixture during the manufacturing process.

- **Toxicity** - Recycled rubber tire material has been found to be nontoxic under leachate testing. However, additives, such as binders, retardants, coatings, and coloring, included in the mix to form and enhance the material, can create potential toxicity problems. These additives are, in some cases, proprietary with the specific formulations kept in confidence by the manufacturer.

- **Structural Strength** - Rubber material, on its own, does not have sufficient rigidity to be considered as a structural component of a noise barrier panel. Therefore, bonding agents must provide adequate stiffness to enable the panels to be considered strong enough to withstand wind loading, or the rubber material must be firmly attached to a suitable stiffener, such as channel backings, cores, or casings.

- **Binders** - Rubber and some binders tend to oxidize over time when exposed to the elements. They may also be susceptible to certain chemical or petroleum products. This increases the potential of premature disintegration of the panels. If concrete is used as a binder, concrete modifiers and special treatment of the crumb rubber are required before they will bond properly to each other. This is particularly important when these panels are exposed to salt, cold weather, and flexing for a long period of time.

- **Coatings** - Some coatings suitable for rubber have a questionable life expectancy. They have a tendency to oxidize prematurely, particularly when used in conjunction with certain pigments. If the surfaces of the noise barrier panels are being manufactured to be sound absorptive, the coatings may clog the surface openings thereby reducing the Noise Reduction Coefficient (NRC).

- **Sound Transmission Class (STC)** - Although the weight of the panels may be sufficient to meet general requirements for minimum STC ratings, it may not be sufficient when produced as a porous panel. Even when stiff backers or cores are used, the nature of this material may require the cores or backers to be extensively perforated to promote bonding.

- **Recyclability** - The recyclability of the final product may have been reduced drastically by the type of additives needed to alter the physical properties of the panel so that it can meet the various fundamental requirements for an effective, safe, and durable noise barrier product.

5.7.2 Verification of Quality.
The tests discussed within this section are described in detail in Section 10 which discusses product evaluation of all types of barrier materials.

- **Flame Retardants** - To ensure that the retardants are adequate, the minimum allowable rate of flame spread and smoke generated should not be greater than the rate for a typical fence material, such as pine.

- **Toxicity** - Concerns for environmental damage and health hazards should be addressed by requesting leachate testing or other methods to determine the toxicity of the final noise barrier panel material.

- **Structural Strength** - The structural strength of the panel must be verified through load testing on a production panel.

- **Bonding** - To optimize the bond between the rubber crumb particles, it is necessary to ensure that the rubber crumb is new or has been protected from the elements. The binders used should be stable under prolonged exposure to ultraviolet light. The manufacturing process should ensure that each rubber particle is completely encapsulated by the binder. If cement is used, the rubber surface should be treated or impregnated with a bonding agent compatible with both the rubber and the concrete. Or, the concrete should contain modifiers that will allow it to firmly bond to the rubber and be able to stand the test of time.

- **Coatings** - The coated panels should be subjected to **weatherometer testing** to determine the longevity of the coating.

- **Noise Reduction Coefficient (NRC)** - If the panels are to be coated, the NRC rating should be verified after the panels have been coated.

- **Sound Transmission Class (STC)** - The assembled noise barrier system should be tested to verify the STC rating. Even though the mass requirement for a suitable STC has been theoretically met, the finished panel may be two porous to actually achieve the desired STC.

5.8 Composites
Composite noise barrier materials, in general terms, can be defined as any product composed of two or more primary materials, such as plywood with a fiberglass skin (see Figure 86), or wood mixed with concrete and then layered onto concrete (see Figure 87). Since the possibilities are almost endless, this section will mainly focus on the special considerations which should be used in evaluating their safety, durability, and performance.

Figure 86. Composite noise barrier (data base #132).

Figure 87. Composite noise barrier (data base #707).

5.8.1 Special Considerations. The combining of basic materials has a tendency to change the performance, durability, and, in some cases, the safety characteristics of the final product. These changes should be investigated thoroughly before the composite materials are used in an actual installation.

- **Burning Characteristics** - Some composite materials may have a tendency to burn, or to be severely damaged under certain conditions. The smoke and emissions that may be generated from burning materials might also be toxic. The ash left from any burnt material may also be considered as toxic and will most likely leach into the surrounding soil.

- **Shrinkage** - The shrinkage rate of the primary materials differ significantly and may cause dimensional instability and leave open cracks between joints or promote accelerated creep, warping, or **delamination**.

- **Ultraviolet Protection** - Some products are very sensitive to ultraviolet light and tend to cause rapid deterioration of pigments, surface appearance, and material strength. To avoid this, it is possible to slow down the deterioration process by adding ultraviolet protection into the composition of the material at the time of molding.

- **Creep** - If plastic is part of the composite material, creep should be considered during the design of the barrier system by reducing the amount of strain which the plastic components may be subjected to.

- **Vandalism** - Some materials are particularly susceptible to vandalism from paint, knives, and lighters.

- **Shatter Resistance** - Although most commonly used products are relatively shatter resistant, this characteristic tends to deteriorate over time and the product becomes more brittle and may shatter on impact by flying objects or vehicles. Damaged panels can usually not be repaired by patching. The only option is to replace the damaged sections, thus increasing the cost of repairs and possibly jeopardizing the appearance of the barrier if similarly molded panels are no longer available or are difficult to reproduce at a reasonable cost.

- **Structural Strength** - Some primary materials used in composite panels do not have sufficient rigidity to be considered as a structural component of a noise barrier panel. Therefore bonding agents must provide adequate stiffness to enable the panels to be considered strong enough to withstand wind loading, or the material must be firmly attached to a suitably stiff backing, core, or casing.

- **Binders** - Some binders tend to oxidize over time when exposed to the elements. They may also be susceptible to certain chemical or petroleum products. This increases the potential of premature disintegration of the panels. If concrete is used as a binder, concrete modifiers, and special treatment of the crumb rubber are required before they will bond properly to each other. This is particularly important when these panels are exposed to salt, cold weather, and flexing for a long period of time.

- **Coatings** - Many coatings have a questionable life expectancy. They have a tendency to oxidize prematurely, particularly when used in conjunction with certain pigments. If the surface of the noise barrier panels are being manufactured to be sound absorptive, the coatings may clog the surface openings thereby reducing the Noise Reduction Coefficient (NRC).

- **Sound Transmission Class (STC)** - Although the weight of the panels may be sufficient to meet general requirements for minimum STC ratings, it may not be sufficient when produced as a porous panel. Even when stiff backers or cores are used, the nature of some materials may require the cores or backers to be extensively perforated to promote bonding.

- **Recyclability** - The recyclability of the final product may have been reduced drastically by the type of additives needed to alter the physical properties of the panel so that it can meet the various fundamental requirements for an effective, safe, and durable noise barrier product.

- **Future Disposal** - The nature of the primary materials or when combined with other materials may render the final product unsuitable for future disposal in land fill sites.

- **Safety** - Consideration to safety issues, such as shatter resistance, should be given when panels are mounted on traffic barriers, such as Jersey barriers.

5.8.2 Verification of Quality.

Some combinations may not have sufficient in-field performance history to be able to determine the long term durability, safety, and performance of specific composites. Therefore testing is much more critical for these types of materials than most others used for noise barrier panels. There are countless standard test methods published to assist in the verification of various materials. Each of which are normally only relevant to very specific material formulations. This handbook will not attempt to list or describe these and suggests current testing information be obtained from other sources. The tests discussed within this section are described in detail in Section 10 which discusses product evaluation of all types of barrier materials. However, the following fundamental tests should be considered for all of these types of materials.

- **Dimensions** - The panel's profile, size, and thickness should be verified since any deviation from design specifications will effect the structural strength, durability, and performance of the noise barrier system (see Section 11.5.1).

- **Flame Retardants** - To ensure that the inherent, or added, flame retardants are adequate, the minimum allowable rate of flame spread and smoke generated should not be greater than the rate for a typical fence material, such as pine.

- **Toxicity** - Concerns for environmental damage and health hazards should be addressed by requesting leachate testing or other methods to determine the toxicity of the final noise barrier panel material.

- **Structural Strength** - The structural strength of the panel must be verified through load testing on a production panel.

- **Bonding** - To optimize the bond between the composites, it is necessary to ensure that the primary materials and binders used are stable under prolonged exposure to ultraviolet light and that the proper binders are used for the specific materials.

- **Coatings** - The coated panels should be subjected to weatherometer testing to determine the longevity of the coating.

- **Noise Reduction Coefficient (NRC)** - If the panels are to be coated, the NRC rating should be verified after the panels have been coated.

- **Sound Transmission Class (STC)** - The assembled noise barrier system should be tested to verify the STC rating.

- **Freeze-Thaw/Salt Scaling** - This test is a combination of two tests, which determines the material's resistance to salt scaling and also to frequent freezing and thawing cycles. It provides a good indicator as to how the final material combination(s) will perform under harsh weather conditions.

5.9 Barrier Surface Treatment

This section describes various surface treatments, including textures, colors, and coatings which may be applied to a noise wall.

5.9.1 Textures.

A vast majority of surface textures are available to the noise barrier designer (see Figure 88). In many cases, such treatments can be applied to several elements of the barrier systems (e.g., posts, panels, caps, etc.). Different barrier surface treatments can be obtained by having different combinations of treatments on these separate barrier elements. The intent of the following discussion is not to address all possible combinations of treatment, but rather, to discuss the more common surface texture types available. The following discussion of texture types is categorized by material type.

Figure 88. Barrier surface treatment: textures (data base #512).

5.9.1.1 Concrete.

- **Smooth Surface** - Such a surface is produced by traditional concrete finishing techniques (see Figure 89). The top side of precast panels cast in a horizontal precast bed are typically "floated" to a smooth texture. Obtaining a smooth surface on the bottom side of such a panel or on either side of a vertically formed panel requires finishing after the initial concrete cure, including filling of voids and final "rubbing" of the concrete with a thin cement mixture.

Figure 89. Concrete: smooth surface (data base #996).

- **Exposed Aggregate** - A stone aggregate surface is typically obtained using precast concrete barrier elements (see Figure 90). The surface is obtained using the selected type, color, and gradation of aggregate in the barrier's concrete mix itself. Care must be taken to ensure that the aggregate selected for its aesthetic appearance also meets the required structural requirements related to strength, angular size, shape, etc. Also, the aggregate should be sufficiently screened, graded, and inspected to ensure the removal of any iron ore aggregate which could give the appearance of rust bleeding from the panel. Prior to the panel being cast, a retarding chemical is placed on the

Figure 90. Concrete: exposed aggregate (data base #748).

form work of the surface which will ultimately have the exposed aggregate finish. The retarder extends the curing time of the concrete on the surface to which it is applied. Different grades of retarder are available which provide varying degrees of penetration into the concrete surface and thus will provide varying degrees of exposure of the aggregate. Following the initial cure of the panel to a degree of strength which enables it to be lifted out of the horizontal form to a vertical position, the surface treated with retarder is power washed with a high pressure (2,000 psi) water wash. This process removes the "retarded" (soft) concrete and exposes the aggregate. An acceptable exposed aggregate surface is most easily obtained on the bottom (down side of a precast panel). Obtaining an acceptable exposed aggregate surface on the top side of a panel is significantly more difficult and may require seeding of the top surface with additional aggregate in addition to the application of a retarder directly to the top panel surface immediately after its pour. Producing consistently acceptable exposed aggregate panels is an art requiring experience on the part of the precaster as well as an emphasis on quality control.

- **Form Liners** - A wide variety of surface treatments can be obtained by the use of form liners (see Figure 91 to 93). Form liners are essentially sheets of material (usually a flexible material such as rubber, but also may be made of wood, metal, or other material) which have been fabricated or molded with one flat side and one side containing the "mold" of the desired surface treatment. These sheets are made to be re-useable for multiple applications. While form liner treatments (related to noise barriers) are most often applied on the down side of panels cast in a precast plant, they can be applied to one or both sides of a panel if cast in a vertical position. Recent innovations have produced machinery capable of pressing a form liner onto the top side of a panel, thus allowing form liner finishes on both sides.

Figure 91. Concrete: form liner (data base #1180).

Figure 92. Concrete: form liner (data base #498).

Figure 93. Concrete: form liner (data base #698).

Care must be taken when such form liner sheets are butted together such as in long precast bays where bulkheads may be moved to produce the required lengths of individual noise barrier panels. In such an application, the joints between form liners will occur mid-panel on many of the noise panels (see Figure 94). To avoid unsightly joint lines which can interrupt the desired surface texture, care must be taken to ensure that the joints are tight, adequately secured to the form work, and "lined up" in terms of the form liner pattern. The potential for such joint lines is greatest with form liners having slight relief (depth of pattern). A deep relief pattern such as a deep rib texture can more easily hide such joints, assuming that the rib patterns are carefully lined up at the joints. Since such form liners by nature create non-

Figure 94. Concrete: form liner panel joints (data base #7029).

smooth surfaces, care must be taken to avoid jagged edges at the ends and sides of the panels. This is particularly critical if the concrete will be painted or stained since any jagged edges can more easily be chipped or broken off, exposing areas of unpainted or unstained concrete.

- **Raked, Broomed or Other Applied Finishes**
  These types of treatments are applicable to the top side of precast panels (see Figure 95). They can result in various degrees of patterns created by hand-applied techniques, with less or equal effort as compared to "floating" a smooth concrete surface.

- **Stamped Finish** - It is possible, with specialized techniques, to press or stamp a design into the top surface of a horizontally poured panel (see Figures 96 and 97). Such techniques have been used (in conjunction with the use of a pigmented surface layer of cement) to create a brick-type surface. Such a treatment is somewhat more labor-intensive than other treatments, particularly if a grout-type look is desired. Any stamping process also requires that the aggregate in the panel be sufficiently deep to allow such stamping.

Figure 95. Concrete: raked finish (data base #508).

Figure 96. Concrete: stamped finish (data base #6512).

Figure 97. Concrete: stamped finish (data base #6514).

- **Inserts** - Special and unique aesthetic treatments can be obtained by the use of panel inserts (see Figures 98 and 99). These inserts are typically precast or manufactured separately from the concrete panel and either imbedded in the panel during the pouring process or set into an equal size and shape indentation within the precast panel. In either case, care must be taken to ensure an adequate bond/attachment between the insert and the noise barrier panel. Such bonds are best accomplished by mechanical attachments (studs or anchors) although chemical bonding techniques may also be employed.

Figure 98. Concrete: inserts (data base #6532).

Figure 99. Concrete: inserts (data base #8015).

- **Veneers** - In the context of this discussion, veneers are meant to represent a separately manufactured material applied to the surface of a concrete noise barrier. Such materials are usually applied for aesthetic reasons although certain veneers may be applied to make the surface sound absorptive. Examples of such veneers include full width and thin brick, ceramic tile, and porous composite sound absorptive materials. The primary concern of the use of veneers is related to assuring an adequate attachment bond as discussed above.

- **Stucco** - Stucco is a finish coat of cementitious material which can be textured in a variety of ways. It bonds directly to the concrete noise barrier without any additional attachments. As such, the cleanliness and roughness of the concrete noise barrier surface is critical in order to ensure an adequate bonding surface for the stucco to avoid spalling (see Figure 100).

5.9.1.2 Masonry Block.
- **Exposed Aggregate** - Compared to the variety of exposed aggregate textures available in a concrete panel, such treatments with masonry blocks are significantly limited. This is due to the limited range of aggregate used in the mass production of concrete blocks.

- **Form Liners** - While some molds may be applied to the production of concrete blocks, their use is rare and limited.

Figure 100. Concrete: stucco (data base #1066).

- **Veneers** - Veneers can be applied to concrete block in a manner similar to that of concrete panels.

- **Fractured Fin** - A common means of achieving a rough, textured surface in concrete block is through the use of fractured fin surfaced block (see Figure 101). This surface is achieved by mechanically sheering the block to create a rough surface.

- **Stucco** - Stucco can be applied to concrete block in a manner similar to that of concrete panels (see Figure 102).

Figure 101. Masonry block: fractured fin (data base #948).

Figure 102. Masonry block: stucco (data base #1061).

5.9.1.3 Brick.

- **Type of Brick** - Literally hundreds of types of brick are available for use in the construction of noise barriers (see Figures 103 and 104). These include more common types of standard brick, including pavers, plus the slump stone and adobe style of brick. Bricks may be laid up with mortar in multiple courses or used to face concrete or concrete block walls as discussed above. Whether bonded to other brick courses or to concrete or block walls, bonding straps of some form are used to secure the system components to each other.

Figure 103. Brick: surface texture (data base #560).

Figure 104. Brick: surface texture (data base #6518).

- **Type of Mortar** - Various colors and types of mortar are available for bonding bricks.

- **Type of Bond** - Similarly, various styles or patterns (known as bonds) are used to construct brick or brick-faced barriers. The most common are stacked bond and running bond. Bricks of different widths, styles, and colors may also be employed to create unique and interesting patterns and/or to create designs representative of area landmarks, themes, etc.

5.9.1.4 Metal.

- **Mechanically Formed Shapes** - Most textures created in metal noise barriers are via the use of roll-formed mechanical brakes, benders, and similar factory-operated devices (see Figure 105). A variety of such formed shapes are available for both preassembled components as well as for components assembled in the field. Although different styles and shapes are possible on each side of the noise barrier (through the use of double-faced panels), surface treatments are most often similar on both sides.

Figure 105. Metal: surface texture (data base #1708).

- **Pressed (dimpled) surfaces** - Smaller relief impressions in metal panels may also be obtained by plant-applied processes.

5.9.1.5 Wood.

- **Plank Orientation** - Different visual appearances can be obtained via the orientation of wood planks used in noise barrier construction. Horizontal, vertical, and diagonal configurations have been employed (see Figures 106 and 107). Planks of different widths and depth can also create interesting visual effects.

Figure 106. Wood: horizontal plank orientation (data base #745).

Figure 107. Wood: vertical plank orientation (data base #730).

- **Battens** - As well as providing a mechanism for preassembling panels on the ground (prior to their attachment to posts) battens can also be employed to create patterns in the noise wall design (see Figure 108).

- **Grain** - Selection of wood grain and roughness can also create the desired surface texture treatment.

- **Lamination** - Various design patterns are possible with laminated panels through the particular orientation of the laminated component elements (see Figure 109). Designs with such a system will most often be similar on both sides of the barrier.

- **Post Type and Orientation** - Wood posts of various shapes and sizes have been employed in the construction of noise barriers. Circular, square, and rectangular post sections are common (see Figures 110 and 111). The exposure or concealment of the post can create different textural and shading patterns.

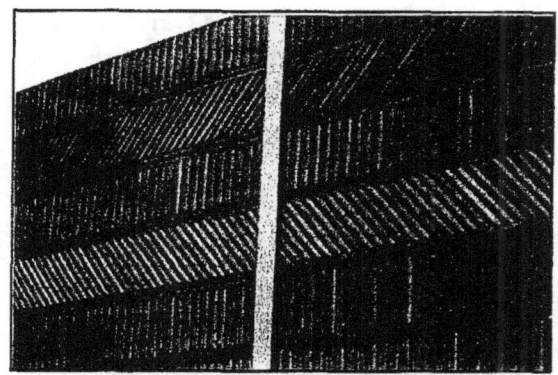

Figure 108. Wood: patterns on battens attached to panels (data base #471).

Figure 109. Wood: pattern on laminated panels (data base #663).

Figure 110. Wood: circular post type (data base #744).

Figure 111. Wood: square post type (data base #435).

5.9.1.6 Transparent Materials.
Surface texture treatments are limited to aesthetic type designs (stencils) applied to such barriers (see Figure 112).

5.9.1.7 Plastics.
Surface textures of plastic panels are generally limited to shapes and textures which can be accomplished via the panel component molding process (see Figure 113).

Figure 112. Transparent: surface stencil design (data base #1954).

5.9.1.8 Rubber.
Similarly, barriers constructed of recycled rubber materials are limited to shapes obtainable through molding of their components. The surface texture of such panels is also influenced to some degree by the density and porosity of the rubber (see Figures 114 and 115).

Figure 113. Plastic: surface texture (data base #792).

Figure 114. Rubber: surface texture (data base #2948).

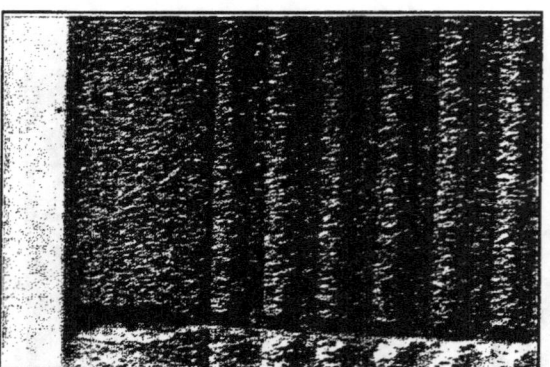

Figure 115. Rubber: surface texture (data base #2949).

5.9.1.9 Composites. Composite panel texture treatment opportunities reflect the availabilities and constraints of the particular components used to form the outside barrier face (see Figures 116 and 117).

Figure 116. Composite: surface texture (data base #136).

Figure 117. Composite: surface texture (data base #708).

5.9.1.10 Other Applications.
- **Planted Walls** - Textures of planted walls are limited to what can be accomplished via the planted vegetation within the wall system.

- **Gunite on Reinforced Chain Link Fence** - A unique texture treatment has been successfully used in a noise barrier application for over twenty-five years. This system consists of a chain link fence which was reinforced with expanded metal mesh and then sprayed with gunite to provide the appearance of a stucco wall (see Figure 118). While such a treatment is probably not advisable close to travel lanes, where it is more prone to damage from airborne debris or vehicular hits, it appears to be a viable and economical technique for use in areas somewhat removed from active travel lanes. The particular wall shown has also withstood several major and many minor earthquakes with little or no damage.

Figure 118. Gunite: surface texture (data base #2243).

5.9.1.11 Special Considerations. As discussed briefly in the introduction of Section 5, the selection of a particular surface treatment texture depends on a number of factors including aesthetic requirements of both sides of the barrier, constructability issues, maintenance concerns, and the type of barrier material. The selection of a form liner finish on both sides of a barrier requires specialized equipment, could negate the ability to use horizontally cast precast barrier elements and could require the use of either vertically cast precast elements or cast-in-place barriers. The inability to use full height precast panels (in a situation where placing such panels is restricted due to overhead wires or other factors) could limit the barrier type, material, and therefore the surface texture options. Texture treatments used with stacked panels should be coordinated so that the joints are either concealed by the pattern or become a part of the pattern (see Figures 119 and 120).

Figure 119. Surface texture: special considerations - stacked panel joint (data base #902).

Figure 120. Surface texture: special considerations - form liner joint (data base #533).

5.9.2 Color. The desired color of noise barriers is provided by one of the following two general techniques or a combination thereof:

- by the natural color of the noise barrier material being used (possibly enhanced by a clear coating)
- by application of a paint, stain, pigmented coating, or integral pigment added to the noise barrier material.

Options related to colors of noise barriers are discussed below for various types of barrier materials.

5.9.2.1 Concrete and Masonry Block.

These types of materials are quite versatile in their ability to be colored (see Figures 121 to 123). The color of their natural elements (sand, stone, and cement) can be varied to obtain different earth tone colors. Addition of a pigment to such a coating can provide a uniform color to a plain concrete surface wall. Pigments can also be added to the mix prior to the forming of barrier components. Quality control and consistency are critical in the production of pigmented concrete since unevenness and blotchiness can become apparent, particularly on smooth surfaces. A consistent color from panel to panel is an important aesthetic factor in achieving a successful barrier system. To ensure a panel-to-panel color consistency, a surface-applied

Figure 121. Concrete: color (data base #2337).

stain may be more effective than the use of integral colors or pigments in the concrete mix. Natural and pigmented barriers have the advantage of not showing damaged areas (from chips, scrapes, etc.) as much as painted or stained surfaces. The maximum protection against the visual effects of such damage can be provided by use of a pigmented panel with a surface stain of a matching color. This reduces any unevenness in the pigmenting process while providing for consistent color throughout the panel. The added cost of such a dual treatment may, however, not be warranted in many cases.

Figure 122. Concrete: color (data base #1218).

Figure 123. Masonry block: color (data base #2373).

5.9.2.2 Brick.
While many variations in color are available with these types of materials, the color is limited to the that of the material itself (see Figure 124).

Figure 124. Brick: color (data base #560).

5.9.2.3 Metal.
Except in cases where a natural (e.g., aluminum, stainless steel, etc.) or rusted look (e.g., weathering steel) appearance is desired, color on metal panels is usually obtained by painting or application of some bonded type of surface coating (see Figure 125). In many cases, barrier elements (particularly posts) are designed with their protective galvanized finish as the ultimate color. This process protects the post while negating any maintenance requirements associated with paint. When colors are desired on metal barriers they may be either applied at the origin of manufacture or in the field. If applied where the barriers are manufactured, such coatings may be applied using a variety of techniques such as a baked enamel finish, a sprayed-on finish, by using a plastisol (a poured-on PVC emulsion), or via a bonded powder coating

Figure 125. Metal: color (data base #1720).

applied via an electrostatic process. In any coating application process, surface preparation is critical. Such preparation needs to consider the initial, intermediate, and final coats in terms of their undercoating and surface preparation requirements. Manufacturers' requirements (related to both coating and material coated) such as extent of sand blasting (white blast, grey blast, etc.), cleaning materials, temperature and moisture controls, etc. need to be adhered to closely. Certain coatings may also allow for easier removal of graffiti without any detrimental effects to panel aesthetics.

5.9.2.4 Wood.
A wide variety of natural colors is available in wood products (see Figure 126). The selection of a particular wood species for its color attributes must also consider other factors of the wood. While a particular wood species may have the exact natural color desired, its qualities related to other areas of concern (durability, warping, rot resistance, etc.) may be unacceptable. Most woods can be stained or painted to obtain a desired color. Clear and pigmented preservative treatments are also available. The type of treatment, stain, or paint must be compatible with the type of wood, nails, and other fasteners used in the barrier assembly, and with the environment in which the barrier will be placed.

Figure 126. Wood: color (data base #464).

Chemical reactions between steel nails and certain wood preservative treatments have caused these nails to corrode, requiring their replacement with stainless steel fasteners. Moisture content at the time of paint or stain application is also critical. It is essential that a wood barrier be properly seasoned before application of any paint or stain, and that no such protective materials are applied when the barrier is damp. Specific requirements will vary for each coating material, necessitating strict compliance with the specifications of the coating manufacturer.

5.9.2.5 Plastics, Fiberglass, and Acrylics.
Color of these materials is most often obtained through the pigmentation of the emulsifiers used in the molding process of barrier elements (see Figure 127). A scratch coat may also be added for protective purposes.

5.9.2.6 Rubber.
Recycled rubber material generally cannot be pigmented. Rubber can be coated (usually with a more expensive polyurethane coating) to obtain a desired color.

Figure 127. Plastic: color (data base #1728).

5.9.2.7 Composites. Composite panel color treatment opportunities reflect the availabilies and constraints of the particular components used to form the outside barrier face (see Figure 128). Particular concern should be paid to ensure compatibility between the barrier materials (including any glues, attaching devices, etc.) and the applied coating to negate any potential for damaging chemical reactions.

5.9.2.8 Planted Walls. Color in such walls can be obtained by the selection of the appropriate plants. Different colors and patterns can be obtained with changes in season.

Figure 128. Composites: color (data base #2533).

5.9.3 Coatings. The discussion in this section is limited to coatings on concrete and masonry barriers, since coatings on other types of barriers are inherently addressed within the discussions of color found in Section 5.9.2.

Coatings are typically applied to concrete or masonry barriers for protective and/or aesthetic reasons. Protection against the elements (wind, rain, salt spray, ultraviolet light, etc.) and potential vandalism (anti-graffiti) are common reasons for application of such coatings. While coatings can significantly enhance the appearance of a barrier or be its primary aesthetic element, care should be taken in application to ensure positive results. Addition of a clear protective coating can have positive benefits in terms of enhancing the color and brilliance of a concrete exposed-aggregate surface. Conversely, application of the same clear protective coating on a plain concrete barrier can have a negative impact by making the barrier surface look wet and enhancing any
unevenness or blotchiness.

5.9.3.1 Anti-Graffiti Coatings. A number of products are on the market which can provide varying degrees of protection against graffiti. Few, if any, barriers can be made graffiti-proof. Quick removal (within 24 hours) of the graffiti may discourage "artists" who try to return to embellish upon the previous night's work. Rougher surfaced and darker colored barriers and barriers covered with vegetation or vines may also provide more resistance to being "hit" by graffiti artists as compared to light colored, non-planted, and/or smooth-surfaced barrier surfaces. However, there is no guarantee that a barrier will not be susceptible to graffiti. Once hit with graffiti, an unprotected concrete or masonry surface can be marred for life due to staining and penetration of the graffiti paint into the concrete substrate, even if the graffiti is "removed." Anti-graffiti coatings are meant to prevent such penetration and to make the removal of the graffiti somewhat easier. Anti graffiti coatings may be clear or pigmented. Clear coatings are usually applied to architectural surfaces such as exposed aggregate or brick (see Figure 129), where the source of the barrier's color is in the natural barrier material itself. In some instances, such coatings applied to exposed-aggregate surfaces tended to enhance its appearance by deepening the aggregate colors. Pigmented coatings are typically applied to natural (unpainted, unstained, un-pigmented) concrete surfaces. Application of coatings is usually done by spraying, although rolling and brushing may occasionally be performed, especially in areas where over-spray is a concern. Anti-graffiti coatings may be of either the permanent or sacrificial variety as described below:

- **Permanent Type** - This type of treatment is aimed at providing a coating which will enable multiple removals of graffiti by high pressure water wash and/or chemical washing techniques without having to replace the coating. Such coatings may be either urethane or water based and usually require a two- or three-coat application process.

Figure 129. Anti-graffiti coating (data base #5332).

- **Sacrificial Type** - This treatment type provides a coating which itself is wholly or partially removed along with the graffiti during the cleaning process. Certain sacrificial coatings may be capable of several cleanings before they must be reapplied. These types of coatings are typically one-coat, water-based or wax-type systems. They are generally less expensive, but do require re-application following a certain degree of graffiti removal.

5.9.3.2 Stains.
In addition to providing the desired color in a noise barrier, a concrete stain can provide a degree of protection against the elements as well as be reapplied to areas "tagged" with graffiti. Stains can be either oil-based or water-based, with the latter being more widely used for concrete applications.

5.9.3.3 Application Process.
Coating of precast barrier elements may be performed at the point of their manufacture (normally a precast plant) prior to their erection or in the field after their erection. Only on rare occasions would coating of precast barrier elements be performed in the field prior to the erection of the barrier. Coating of cast-in-place noise barriers must occur in the field. If coating is done at the plant, conditions may be better controlled, but barrier elements must be stored during drying or between coats at a location free of dust and potential damage from plant operations. Additional care must also be taken during transport of coated panels to the job site and during their erection process to protect them against possible damage. It is for these reasons that the majority of coating is performed in the field following the erection of the barrier. Even in this situation, coating must be done when conditions (rain, temperature, wind) are acceptable.

5.9.3.4 Relationship of Coating Type to Maintenance Philosophy.
Anti-graffiti coatings should only be applied if the responsible organization has a policy which dictates the removal of graffiti from barriers. If the standard operating procedure related to the treatment of graffiti is to paint

over it, then anti-graffiti coating is not only a waste of money, but also creates a surface to which the "paint over" paint will not easily adhere. With this maintenance philosophy, the use of stains appears to be more consistent and cost-effective.

5.9.3.5 Relationship of Coating to Barrier's Acoustical Performance.

In addition to the factors listed above, it is essential that the coating be compatible with the intended acoustical performance of the barrier. If the barrier is designed to be sound reflective (see Section 3.5.4), then any type of coating may be applied without affecting its acoustical performance. However, if the barrier is to perform a sound absorptive function, any coating used must not interfere with the barrier's sound absorptive characteristics. For instance, coating a sound absorptive, porous surface with a paint or a urethane-based anti-graffiti coating could seal up the voids which provide the barrier's sound absorptive characteristics. A water-based penetrating stain may be an appropriate coating in this instance. Before applying any coating to a sound-absorptive surface, research and coordination with both the barrier manufacturer and the coating supplier is strongly recommended.

5.9.3.6 Health and Environmental Issues.

While significant benefits in several areas can be realized by the use of coatings, there are health and environmental issues which need to be addressed in association with their use. As a general rule, products should be stored, applied, and disposed of in strict compliance with the recommendations of the manufacturer and in accordance with all applicable federal, state, and local regulations. Urethane-based coatings require particular attention due to their higher content of volatile organic compounds (VOC) and their more potent odors. Water-based and other more recently developed coatings will have little or no VOCs and less of an odor problem. They also have less toxicity. Over spray is still a factor requiring careful consideration where barriers are in close proximity to either residences or vehicles or where coating application is being performed near water courses or swales leading to such sources of water. Adequate covering of the ground with tarpaulins can help to minimize ground and water contamination.

Section Summary

Item #	Main Topic	Sub-Topic	Consideration	See Also Section	✓
5-1	Concrete	Aesthetic	For cast-in-place: form liners and architectural inserts must be placed on vertical surfaces of the form work which can increase the chance of imperfections in the wall surface.	5.1	
			For cast-in-place: application of concrete-retarding chemicals to the vertical form work surfaces for the purposes of obtaining an exposed aggregate finish is difficult.	5.1	
			For cast-in-place: other surface textures obtained through raking, brushing, or stamping of concrete are not possible.	5.1	
			Surface textures: ■ Smooth Surface - Ensure a smooth surface by performing a final "rubbing" of the concrete with a thin cement mixture. ■ Exposed Aggregate - Provide sufficient screening, grading, and inspection to ensure the removal of any iron ore aggregate which could give the appearance of rust "bleeding" from the panel. ■ Form Liners - The joints must be tight, adequately secured to the form work, and "lined up" in terms of the form liner pattern. Care must be taken to avoid jagged edges at the ends and sides of the panels. ■ Inserts - Care must be taken to ensure an adequate bond/attachment between the insert and the noise barrier panel. ■ Veneers - The primary consideration of the use of veneers is related to assuring an adequate attachment bond. ■ Stucco - The cleanliness and roughness of the concrete noise barrier surface is critical in order to ensure an adequate bonding surface for the stucco.	5.9.1.1	
			Quality control and consistency is critical in the production of pigmented concrete since unevenness and blotchiness can become apparent, particularly on smooth surfaces. The maximum protection against the visual effects of damage can be provided by use of a pigmented panel with a surface stain of a matching color. The added cost of such a dual treatment may, however, not be warranted in many cases.	5.9.2.1 5.9.3	
		Drainage and Utility	Application of concrete-retarding chemicals to the vertical form work surfaces for the purposes of obtaining an exposed aggregate finish is difficult.	5.1	
		Safety	Consider on-site material testing and inspection during construction.	5.1	
		Installation	For precast, consider size limitations, shipping requirements, traffic implications, reusability of precast panels, quality assurance process	4.1.2.3.1 5.1	
			For cast-in-place, consider on-site material testing and inspection procedures during construction, and weather concerns for on-site casting and curing.	4.1.2.6 5.1	
		Maintenance	For free-standing noise walls, consider access for landscaping.	4.1.2.6	

Item #	Main Topic	Sub-Topic	Consideration	See Also Section	✓
5-2	Brick and Masonry Block	Aesthetic	Surface textures: ■ Exposed Aggregate - Consider limited use due to mass production constraints. ■ Form Liners - While some molds may be applied to the production of concrete blocks, their use is rare and limited. ■ Veneers - Ensure an adequate attachment bond. ■ Stucco - The cleanliness and roughness of the noise barrier surface is critical in order to ensure an adequate bonding surface for the stucco.	5.2 5.9.1.2	
			Quality control and consistency is critical in the production of pigmented blocks since unevenness and blotchiness can become apparent, particularly on smooth surfaces.	5.2 5.9.2.1 5.9.3	
		Structural	Consider the need for a continuous concrete foundation.	5.2 8.4	
			Consider the compressive strength of the concrete materials.	5.2	
		Installation	Hand-laid versus preassembled panels: ■ Consider each type's versatility to conform to ground contours. ■ Consider each type's speed of erection. ■ Consider special leveling courses on grades of up to 6 percent. ■ Consider the scaffolding requirements for hand-laid panels.	5.2 11.1	
5-3	Metal	Acoustical	Consider the Sound Transmission Class requirements.	5.3	
		Aesthetic	Consider the possible "industrial" appearance of metal walls.	5.3	
			Consider weathering steel concerns because unpainted rusting panels can stain adjacent concrete.	5.3	
			Manufacturers' requirements (related to both coating and material coated) such as extent of sand blasting, cleaning materials, temperature and moisture controls, etc. need to be adhered to closely.	5.9.2.3 5.9.3	
		Structural	Consider the non-compatibility of various metal combinations.	5.3	
			Ensure corrosion resistance.	5.3	
			Consider the metal's structural strength.	5.3	
		Safety	Consider the possible glare due to on-coming vehicles.	5.3	
			Consider implementing a deterrent for climbing on barrier girts.	5.3	
		Maintenance	Debris and errant vehicles easily causes noticeable damage.	5.3	
5-4	Wood	Acoustical	Consider possible shrinkage and warping causing noise leakage through gaps.	5.4	
			Ensure a tight fit for tongue and groove planking to avoid noise leakage.	5.4	

Item #	Main Topic	Sub-Topic	Consideration	See Also Section	✓
		Aesthetic	Selection of a wood species for its color attributes must also consider durability, warping, rot resistence, etc. The type of treatment, stain, or paint must be compatible with the type of wood, nails, etc. used in the barrier assembly, and with the environment in which the barrier will be placed. Chemical reactions between steel nails and wood preservative treatments may corrode nails. Moisture content at the time of paint or stain application is also critical.	5.9.2.4 5.9.3	
		Safety	Consider the wood's burning characteristics when choosing a wood.	5.4	
			Consideration should be given to a wood's shatter resistance.	5.4	
5-5	Transparent Panels	Structural	Consider the various methods of mounting.	5.5	
			Consider edge conditioning the panels.	5.5	
		Safety	Consideration should be given to shatter resistance.	5.5	
			Consider the possible glare due to on-coming vehicles.	5.5	
		Maintenance	Consider the methods of cleaning the panels.	5.5	
			Consider the maintenance concerns related to vandalism and scratches.	5.5	
			Consider the need for ultraviolet light protection.	5.5	
			Debris and errant vehicles easily causes noticeable damage.	5.5	
		Cost	Transparent noise barriers costs can be more costly than common concrete or steel panels.	5.5	
5-6	Plastics	Acoustical	Consider the possible shrinkage in plastic materials.	5.6	
		Safety	Consideration should be given to the material's shatter resistance.	5.6	
			Consider the possible glare due to on-coming vehicles.	5.6	
			Consider the burning characteristics of the materials.	5.6	
		Maintenance	Consider the need for ultraviolet light protection.	5.6	
			Consider the maintenance concerns related to vandalism and scratches.	5.6	
5-7	Recycled Rubber	Acoustical	Consider possible coating interference with the material's Noise Reduction Coefficient.	5.7	
			Ensure an adequate Sound Transmission Class.	5.7	
			Recycled rubber material generally cannot be pigmented. Rubber can be coated (usually with a more expensive polyurethane coating) to obtain a desired color.	5.9.2.6 5.9.3	
		Structural	Consider the panel's structural strength requirements.	5.7	
			Consider the material's bonding requirements.	5.7	
		Safety	Consider the need for flame retardants.	5.7	
			Consider the material's possible toxicity concerns.	5.7	
5-8	Composites	N/A	(Refer also to concerns for individual materials within composites.) Particular concern must be paid to ensure compatibility between the barrier materials (including any glues, attaching devices, etc.) and the applied coating to negate any potential for damaging chemical reactions.	5.8 5.9.3	

6. NOISE BARRIER AESTHETICS

Aesthetics is an issue that should be of concern to all people involved in the ultimate selection and design of a noise barrier. It is often felt to be as important as the noise reduction provided by the barrier and is the most subjective of any aspect of noise barrier design, with the phrase "beauty is in the eye of the beholder" often used in discussing noise barrier aesthetic treatments. Whether a jagged, stepped, sloped, uniform, non-uniform, colored, plain, straight, curved, or textured barrier is desired at any given location is a decision left to the responsible organization based on its policies and procedures regarding design philosophies, community input, and any other factors which are considered in the decision-making process related to barrier aesthetics. Public input should always by considered in the aesthetic design of noise barriers. The intent of this section is not to justify any particular philosophy related to any element of aesthetic design, but rather to discuss elements that should be considered regardless of the particular aesthetic philosophy chosen.

6.1 Relationship of Noise Barrier to Surroundings

In designing noise barriers, there are two general approaches or philosophies related to aesthetic treatments. One philosophy is to aesthetically design the noise wall in a manner that it blends into the surrounding environment and is as unintrusive as possible. The other philosophy is to have the noise barrier be a prominent feature in the surrounding environment. Neither should be considered right or wrong in a general sense. Both philosophies have been successfully employed and even combined on the same project. In certain instances, highway sides of noise barriers have incorporated a "blend in" philosophy while community sides of the same barriers have employed more prominent architectural treatments. Certain elements of aesthetic design should be evaluated and considered separately in the design process dependent upon whether the barrier surface is being seen from the highway or from its adjacent land uses.

Prior to discussing aesthetic design issues specific to the views of the motorist (see Section 6.1.6) and the community (see Section 6.1.7), a number of aesthetic design issues common to both barrier view points are described below.

6.1.1 Alignment Changes.
While on occasion, a barrier can be constructed at a continuously uniform distance from the roadway and at a uniform height or elevation, it is rare that barriers can be built without some change in horizontal and vertical alignment. In attempting to make aesthetically pleasing barrier transitions and profiles, barrier designers incorporate shifts and transitions into the barrier's alignment. Such changes must be made within the restrictions and tolerances of the barrier system components. For example, angles of horizontal alignment shifts on post and panel systems are restricted to those which the particular post design can accommodate. Barrier systems with cable-secured, linear ball and socket-style panel connections can accommodate much greater angles.

Combined shifts in both horizontal and vertical alignment (see Figure 130) can create conditions which may not be obvious to the noise barrier designer unless the barrier can be viewed from various angles. Such conditions can occur in areas where a barrier transitions from a location on the edge of **shoulder** of a fill section to a point at the top of a cut section (see Figure 131). It should also be noted that the horizontal angle of the back (community) side of the barrier's transition section can actually reflect flanking sound waves back into the community which the barrier is designed to protect (see Figure 132). While such a condition cannot always be avoided, its recognition during the design process can enable the adverse condition to be rectified by placing acoustically absorptive material on the normally reflective back side.

Figure 130. Alignment changes (data base #1496).

Figure 131. Alignment changes (data base #6524).

Figure 132. Alignment changes: possible flanking reflections (data base #8030).

6.1.2 Vertical Stepping/Sloping of Panels.

Depending upon the type of barrier system utilized, vertical transitions in noise barriers can be accomplished in a variety of manners. Such transitions in post and panel systems are often accomplished by stepping the panels. A uniform appearance can be provided by designing barriers with sections containing consistently spaced equal height steps (see Figure 133). An irregular appearance can be provided by providing random height steps at irregular intervals (see Figure 134). To avoid having to cast non-rectangular panels, and for aesthetic reasons, such steps normally are made at the location of the posts. Keep in mind that on radically changing terrain, consideration should be given

Figure 133. Vertical stepping of panels: uniform (data base #6523).

to sloping the bottom of the panels to avoid burying a large portion of the panels in the ground (see Figure 135). This would avoid reducing panel lengths (to ensure structural stability) and decreasing the distance between posts which would increase the number of posts required and the costs for more posts and foundations. Barrier transitions can also be accomplished using a smooth sloped top of barrier profile (see Figure 136). This technique is common with cast-in-place noise barriers. If this technique is used in a post and panel system, irregularly shaped panels are required, and consideration should be given to also sloping the post tops at a consistent angle.

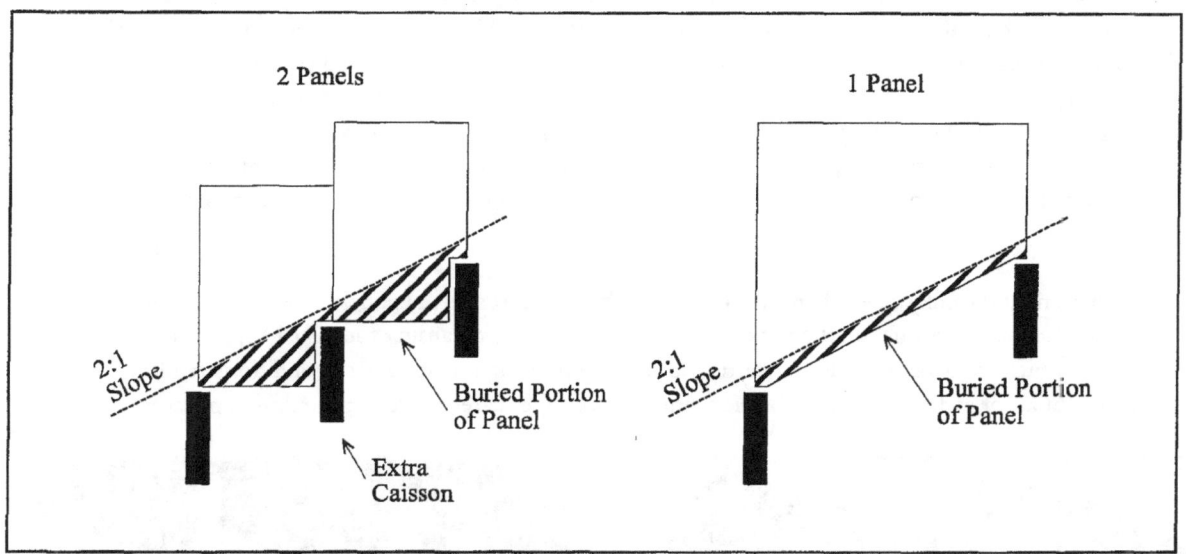

Figure 134. Vertical sloping of panels.

Figure 135. Vertical stepping of panels: irregular (data base #111).

Figure 136. Vertical sloping of panels: smooth (data base #1839).

6.1.3 Caps. For the purpose of this discussion, caps are considered to be separate elements of the barrier system applied to either the top of noise walls or to the top of the noise wall posts. The "cap look" is accomplished as an integral part of the fabrication/construction of the noise barrier wall panels.

6.1.3.1 Horizontal Caps. Caps have been placed on the top of noise barriers (panels, posts, or both) for both aesthetic and acoustical reasons (see Figures 137 to 140). Caps can smooth a barrier's profile eliminating saw-toothed steps and gaps and provide a pleasing shading pattern. However, care should be taken to keep the size of the cap proportional to the scale of the noise wall. Too large of a cap can give the visual perception of the noise wall being "top heavy." A cap can also interfere with the natural "washing" of the top portion of the noise wall which occurs during rain events. With the noise wall not being uniformly washed, streaking becomes more apparent over time and can become very unsightly.

Attachment and caulking details need to be carefully considered at the panel-to-post attachment points and between cap sections. Particular concern should be taken regarding the visual appearance of capped barriers which follow a meandering vertical and horizontal alignment. These conditions tend to create the potential for awkward looking barriers unless the proper care is taken in the design process.

Figure 137. Noise wall horizontal cap (data base #271).

Figure 138. Noise wall horizontal cap (data base #1325).

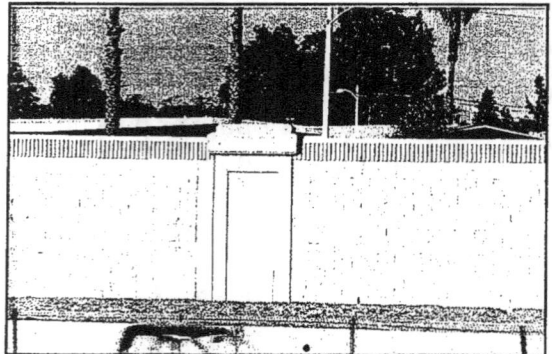

Figure 139. Noise wall horizontal cap (data base #2434).

Figure 140. Noise wall horizontal cap (data base #1731).

6.1.3.2 Vertical Caps. Capping of vertical posts can provide a more aesthetically pleasing barrier system but requires careful considerations in order to avoid adverse maintenance situations (see Figures 141 and 142). Capping of a steel post with a pre-manufactured cap can negate the need to provide a visually pleasing treatment on the steel post itself. However, sufficient treatment of the steel post should be provided to ensure durability and reduce the likelihood of premature rusting. The design of the cap and post should be consistent with long-term maintenance anticipations. For instance, if it is necessary to remove the cap from time to time, the attachment details may be different than if the cap-to-post attachment is considered "permanent." In either case, drainage considerations are critical and should be considered in light of the respective cap and post materials to avoid trapping of water, resulting in premature rusting, warping, or other material degradation.

Figure 141. Noise wall vertical cap (data base #2541).

Figure 142. Noise wall vertical cap: damage (data base #5224).

6.1.4 Barrier End Treatments. Several methods have been successfully used to create aesthetically pleasing treatments at the ends of noise barrier systems. Where topography permits, the barrier end can be buried into the existing ground (see Figure 143). Barriers can also be curved back away from the road at their end points. This technique may have an added advantage of providing some additional acoustical abatement of flanking noise while softening the end of the barrier (see Section 3.5.2). Ends of barriers can be reduced in height (using stepped rectangular panels as in Figure 144, or sloped panels as in Figure 145) from their acoustically required height to a height of approximately 1.5 m (5 ft), equal to right-of-way fence height. While such a treatment may provide the desired aesthetic treatment, it is likely to require construction of some area of barrier which is not absolutely necessary for acoustical reasons. Decisions related to such a treatment should weigh the added costs against the aesthetic benefits and any additional acoustical benefits provided. Ending the barrier at its required acoustical height and buffering its end points with plantings (see Figure 146) and/or berming (see Figure 147) are other techniques.

Figure 143. Barrier end treatment: buried into existing ground (data base #80).

Figure 144. Barrier end treatment: stepped panel (data base #193).

Figure 145. Barrier end treatment: sloped panel (data base #2385).

Figure 146. Barrier end treatment: vegetation (data base #1243).

Figure 147. Barrier end treatment: berming (data base #1270).

6.1.5 Special Aesthetic Considerations in Cultural/Historic Areas. Special barrier aesthetic treatments may be required in areas of cultural and/or historic significance. Often such treatments have been incorporated via special inserts, castings, or designs which reflect the historic and/or cultural characteristics of the community (see Figures 148 and 149).

Figure 148. Special considerations in cultural areas (data base #6533).

Figure 149. Special considerations in historic areas (data base #6516).

6.1.6 View from the Road. The view of noise barriers experienced by drivers and occupants of vehicles traveling on the highway is significantly different from the view experienced by adjacent land users. From a vehicle, a long expanse and wide viewing angle of a barrier can be seen in a very short time period. Small detail elements and textures are, therefore, less apparent from this perspective. The barrier is most often seen by the driver in a series of generally low angle views and its overall shape and patterns (the relationship of different barrier elements) becomes more apparent (see Figures 150 and 151). Issues related to the view of a noise barrier from the driver's perspective are complicated by the fact that the barrier is viewed from a different perspective by drivers traveling in one direction compared to those driving in the opposite direction.

Figure 150. View from the road (data base #2007).

Figure 151. View from the road (data base #3128).

6.1.6.1 Color. The overall color of a barrier viewed from the driver's perspective becomes a major visual element. Depending upon the particular design philosophy, the chosen color can draw the eye towards the barrier (see Figure 152 and 153) or tend to blend it into the background of the surrounding terrain. In settings where trees and natural vegetation form the backdrop for the barrier, neutral to dark earthtone colors can make the barrier less obtrusive, while lighter and non-earthtone colors can make the barrier stand out. When viewed against an open backdrop such as the sky, lighter colored barriers may be less obtrusive.

Figure 152. View from the road: color (data base #3118).

Figure 153. View from the road: color (data base #1218).

6.1.6.2 Texture. For texture treatments on barriers to be noticeable and meaningful from the driver's perspective, they need to have fairly deep patterns and generally should be capable of creating shadow effects within the pattern itself. Aside from instances where textures are applied to create colors (such as exposed aggregate) or to deter graffiti, they provide little benefit if the design philosophy is to blend the wall into its surroundings. They can be a major element in helping to emphasizing a barrier's aesthetics if appropriately coordinated with color and pattern elements (see Figures 155 and 156).

Figure 154. View from the road: texture (data base #2368).

Figure 155. View from the road: texture (data base #652).

6.1.6.3 Pattern. The relationship of different barrier elements (posts, panels, adjacent panels, caps, etc.) is referred to as the barrier's pattern. With the blended barrier philosophy, pattern is often de-emphasized by keeping the color and texture consistent for all barrier elements. On the other hand, the barrier's presence can be emphasized by the use of different patterns. Some examples of the wide variety of techniques used to create patterns include varying the color and/or texture of adjacent panels; providing a different color/texture on posts and/or caps than on panels; and changing the color, relief, and/or texture within the panel itself. On a long stretch of barrier, pattern (such as the occasional introduction of a non-standard panel) can help to break up the monotony of the barrier (see Figures 156 to 162).

Figure 156. View from the road: pattern (data base #1122).

Figure 157. View from the road: pattern (data base #453).

Figure 158. View from the road: pattern (data base #2296).

Figure 159. View from the road: pattern (data base #2316).

Figure 160. View from the road: pattern (data base #2352).

Figure 161. View from the road: pattern (data base #328).

Figure 162. View from the road: pattern (data base #5386).

6.1.6.4 Shape. The shape of a noise barrier is defined by its horizontal (plan view) and vertical (profile view) configurations. A change in either the horizontal geometry or vertical profile of a noise barrier can in itself have dramatic or subtle implications in terms of the aesthetics of the barrier (see Figures 163 and 164). Similarly, the manner (uniform, non-uniform, random) in which changes in plan and elevation occur will result in either a smooth, varied, or jagged barrier shape. Barriers can be designed to meander (in plan view) and follow existing ground contours, thus creating many visually interesting configurations. Such treatments can create shapes which cast shadows, thereby giving the overall barrier a different appearance at different times of the day. Such flexibility can also enable barriers to avoid obstacles (poles, inlets, trees, etc.) that would otherwise have to be relocated or removed.

Figure 163. View from the road: shape (data base #1238).

Figure 164. View from the road: shape (data base #1190).

The most visible portion of the noise barrier in terms of its shape is usually its top, especially when it is viewed against a uniform backdrop such as the sky or a uniform contrasting colored background. It is for this reason that particular attention needs to be paid to the top of a barrier. Due to the types of plans and profiles typically available to individuals developing the final acoustical top of a barrier profile and the final profile in the plans, specifications, and estimate (PS&E) drawings, top of barrier profiles are often developed on drawings viewed at a right angle to the barrier (and typically the highway), and with an exaggerated horizontal scale. While an apparent desired (uniform, jagged, etc.) top of barrier profile may be developed using such plans, the actual profile (as viewed by drivers on the highway) may not meet the intent of the designer. A true profile can only be ensured if one can view the barrier from the true perspective of the drivers (traveling in both directions) and from various locations along the highway. Fortunately, computer-aided drafting techniques and programs, such as the Federal Highway Administration's Traffic Noise Model (FHWA TNM®), enable the designer to evaluate the barrier from such a perspective. Even after such considerations result in an acceptable top of barrier profile, the profile should be reviewed in terms of its relationship to the ground profile along the base of the barrier to ensure that no unplanned awkward relationships exist.

6.1.7 View from the Adjacent Land Uses.

The view of noise barriers experienced by occupants of properties behind the noise barrier (community side) is most often influenced by a relatively small, specific portion of a noise barrier system. Because of the potential closeness of such barriers to their protected receptors, the relative height of the barrier in proportion to the distance from the receptor is a factor requiring consideration. The appearance of a barrier overpowering a protected receptor by creating unwanted shadows (see Figure 165), impeding natural air flows, and/or blocking panoramic views needs to be weighed against the acoustical benefits in any decision-making process. Small detail elements and textures in

Figure 165. View from adjacent land uses (data base #148).

the barrier are more easily seen and therefore are more apparent from this perspective. Since a relatively small section of the barrier is most often seen by any one observer, its overall shape and patterns are less of a factor. In general, the visual dominance of a noise barrier near residences is reduced when the barrier is placed at a distance of at least two to four times the barrier's height. Additional landscaping on the residential side may also help to reduce a barrier's visual impact.[18]

6.1.7.1 Color. The overall color of a barrier viewed from the community perspective is a major visual element and the discussions in Section 6.1.5.1 pertaining to color from the roadway perspective are applicable also to the community side of the barrier (see Figure 166 and 167).

Figure 166. View from adjacent land uses: color (data base #7039).

Figure 167. View from adjacent land uses: color (data base #8069).

6.1.7.2 Texture. Detailed texture treatments on barriers are noticeable and meaningful when viewed from an observer in a stationary position on the community side of a noise barrier (see Figure 168). While deep textures can provide a desired look, textures of lesser relief can be successfully used in environments where the barrier is in relatively close proximity to the receptor. However, they can be a major element in helping to emphasize a barrier's aesthetics if appropriately coordinated with color and pattern elements.

Figure 168. View from adjacent land uses: texture (data base #653).

6.1.7.3 Pattern. As discussed in Section 6.1.5.3, pattern can play a major role in barrier aesthetics (see Figures 169 to 172). In the more confined and closely viewed community side environment, patterns need not be as bold or as large as those required along the highway side. Even if the desired philosophy tends toward uniformity of aesthetics, different community side patterns can be utilized in different areas since in many cases, only a small section of barrier is visible from any one location.

Figure 169. View from adjacent properties: pattern (data base #458).

Figure 170. View from adjacent properties: pattern (data base #2420).

Figure 171. View from adjacent properties: pattern (data base #2562).

Figure 172. View from adjacent properties: pattern (data base #2309).

6.1.7.4 Shape. While much of the discussion related to shape in Section 6.1.5.4 is also pertinent to the community side views, specific details regarding barrier plan and profile are important for the portion of the barrier seen from any particular viewpoint. As such, horizontal shifts and top of barrier steps, slopes, and transitions, while possibly having a minor visual impact from a driver's view, can be significant from a community standpoint (see Figure 173). This is particularly noticeable where a transition (such as a step in the top of a barrier profile) or a horizontal shift occurs in the middle of a specific property. Planning such transitions to occur at property lines can in some cases minimize these types of adverse visual conditions. Since the community side of barriers is viewed from a stationary position and often from an angle perpendicular to the barrier, the need to view the barrier at shallow angles is not as critical as for the highway side.

Figure 173. View from adjacent properties: shape (data base #664).

6.2 Landscaping

6.2.1 Integration of Noise Barrier with Surroundings and Accommodation of Existing Vegetation.
Landscaping in the vicinity of noise barriers should be integrated with the landscaping theme chosen for the general highway environment as well as being compatible with the existing landscaping (if adequate and acceptable) of the adjacent land uses and surroundings (see Figures 174 and 175). This applies whether the noise barrier is a solid wall, a berm, a combination wall and berm, or a planted barrier. Wherever possible, consideration should be given to accommodating existing vegetation in the design process. It is suggested that a field review be conducted with a landscape architect or other knowledgeable tree expert to "flag" significant trees/vegetation to avoid/saved, if practical, before the final wall alignment is set. This dictates a commitment to consider integrating the horizontal alignment of the wall with the existing topography and can have a bearing on the type of noise barrier material, the footing type, and the size of noise barrier components utilized. The vertical profile of the barrier can also be influenced by these factors. A cooperative effort should be made balancing good engineering practice with environmental sensitivity.

Figure 174. Landscaping: integration with existing vegetation (data base #2560).

Figure 175. Landscaping: integration with existing vegetation (data base #6526).

6.2.2 Supplementing Existing Vegetation, Replacing Existing Vegetation, and/or Adding New Vegetation. In areas where the existing landscaping is sparse or not of the type deemed desirable, consideration of supplementing or replacing such vegetation with new plantings should be given. Such plantings can be in the form of trees, bushes, shrubbery, and vines placed in the vicinity of the barrier (see Figures 176 and 177). Various methods have been utilized to plant vines, which ultimately climb the barrier (see Figure 178). One method of creating a vine-covered noise barrier involves drilling angled holes through the noise barrier wall, planting vines behind the walls, and training them to grow through the holes to the highway side (see Figure 179). This method is particularly applicable in areas where space on the highway side is not available for plantings.

Figure 176. Landscaping: supplementing vegetation (data base #1975).

Figure 177. Landscaping: supplementing vegetation (data base #6530).

Figure 178. Landscaping: supplementing vegetation (data base #470).

Figure 179. Landscaping: supplementing vegetation (data base #820).

In areas where space on the highway side is available between a protective barrier (such as a Jersey barrier or steel guard rail) and the noise barrier, this area can be used for planting of vegetation, including vines (see Figures 180 and 181). In the case of a Jersey barrier, a raised planter can be created in the space between the protective barrier and the noise barrier. The type of vegetation capable of being planted and maintained in this area is dependant upon its width, soil type, irrigation (natural or artificial), orientation (full sun, shade, etc.), and climatic conditions. Even a narrow space between the noise wall and the protective barrier may be adequate to support vine growth. Such a treatment can also soften the appearance of the barrier and reduce its apparent height.

Figure 180. Landscaping: supplementing vegetation (data base #51).

Figure 181. Landscaping: supplementing vegetation (data base #1759).

Other specific applications where planting in the vicinity of noise barriers may be appropriate are discussed below along with other planting considerations:

- In the vicinity of steps in the top of a barrier profile. Vegetation, typically trees, can soften or hide such steps and can be particularly useful in areas where large steps are unavoidable;

- At the ends of noise barriers, particularly where barriers cannot be stepped down or curved back;

- In areas known to be susceptible to graffiti. It may be far more cost effective to increase plantings on or in the vicinity of a plain surface barrier than to try to deter graffiti by providing a textured treatment with an anti-graffiti coating; and

- In pockets created by meanders or jogs in the noise barrier.

While a continuous planting scheme along a barrier can be beneficial, it can also become monotonous. Occasionally breaking up this continuous planting scheme with denser plantings can add interest and create diversity. Such diversity can also be obtained by varying the species, colors, and sizes of vegetation.

It is essential that the landscape plan be coordinated with the engineering of the noise barrier and with its aesthetic design. If such coordination does not occur, situations such as the following can occur:

- Plantings screen or block aesthetic features of the noise barrier (see Figure 182). Trees, high scrubs, and vines could hide aesthetic inserts, designs cast in noise barriers, or other specifically designed aesthetic features of the noise barriers;

- Plantings interfere with drainage in the vicinity of the barrier. Drainage under, along, or through the noise barrier could be affected by landscaping placed in inappropriate locations.

- Plantings interfere with maintenance or emergency access features of a particular barrier design. Plantings could restrict access through barrier overlap areas, to access doors or fire hose openings/valves, or to the noise barrier itself. Vines could grow in or around such fire hose valves, interfering with their use. Plantings could also obscure the identification signs for these access features;

Figure 182. Landscaping: blocking panel aesthetic features (data base #1212).

6.2.3 Consistency of Landscape Treatment with Maintenance Philosophy.
No matter how well designed a landscape plan may be from its aesthetic standpoint, it is only as good as the ability of the responsible organization to adequately maintain it. It is a waste of time and money to design an aesthetic treatment for which there is neither the commitment (in terms of manpower), funding (long term) to adequately maintain, or coordination with other maintenance considerations. Figure 183 shows a planted barrier that wasn't adequately watered. Figure 184 shows a barrier with a stain applied around the vine growth causing unstained patches on the wall; the landscapers should have coordinated the timing of their plantings with the maintenance personnel assigned to stain the wall. No matter what the desire from an aesthetic standpoint, the landscape plan needs to be responsive to these constraints. Such constraints may appropriately lead to the selection of vegetation that is native "maintenance free" and to a plan that will foster growth of natural vegetation.

Figure 183. Landscaping: consistency with maintenance philosophy (data base #6531).

Figure 184. Landscaping: consistency with maintenance philosophy (data base #2155).

Section Summary

Item #	Main Topic	Sub-Topic	Consideration	See Also Section	✓
6-1	Alignment Changes	Acoustical	Shifts and transitions into the barrier's alignment must be made within the restrictions and tolerances of the barrier system components. Combined shifts in both horizontal and vertical alignment must avoid reflecting flanking sound waves back into the community.	6.1.1	
6-2	Vertical Stepping/ Sloping of Panels	Aesthetic	To avoid having to cast non-rectangular panels, stepping of panels should be made at the location of the posts with consideration also given to sloping the post tops at a consistent angle.	6.1.2	
6-3	Caps	Aesthetic	Consider the aesthetic concerns related to the size of the cap in proportion to the scale of the noise wall and related to the horizontal and vertical alignment of the cap with the entire barrier.	6.1.3	
		Drainage and Utility	Provide for adequate drainage requirements.	6.1.3	
		Structural	Attachment and caulking details need to be carefully considered at the panel-to-post attachment points and between cap sections.	6.1.3	
		Maintenance	Barriers with large horizontal caps may shade the top portion of a barrier and prevent the natural cleansing of that area by rain water.	6.1.3	
6-4	Barrier Ends	Cost	When considering a barrier end treatment, the decision should weigh costs against any acoustical and/or aesthetic reasons.	6.1.4	
6-5	View from the Road		Small detail elements and textures are less apparent from this perspective. The barrier is seen from low angle views, and its overall shape and patterns become more apparent. Also note the different perspective of drivers traveling in opposite directions.	6.1.6	
6-6	View from Adjacent Land Uses		Because of the potential closeness of barriers, the relative height of the barrier in proportion to the distance from the receptor is a factor requiring consideration. Horizontal shifts, top of barrier steps, slopes, and transitions property boundaries require planning to minimize adverse visual conditions.	6.1.7	
6-7	Landscaping	Aesthetic	Trees, high scrubs, and vines could hide aesthetic inserts, designs cast in noise barriers, or other specifically designed aesthetic features of the noise barriers.	6.1.8	
		Drainage and Utility	Drainage under, along, or through the noise barrier could be affected by landscaping placed in inappropriate locations.	6.1.8	
		Safety	Plantings could restrict access through barrier overlap areas to access doors or fire hose valves, or to the noise barrier itself. Plantings could also obscure the identification signs for these access features.	6.1.8	
		Litter	Landscaping in a high litter area should also consider what type of vegetation is best to use. A thorny type of bush may make litter cleanup more difficult than such litter removal from a grassy area.	12.7	

7. DRAINAGE AND UTILITY CONSIDERATIONS

This section describes the various concerns associated with drainage and **utilities**, such as lighting and sign supports, around noise barriers.

7.1 Drainage Requirements

Because of their locations and relationships to the highway, barriers often interfere with normal drainage patterns related to surface runoff paths both parallel and perpendicular to the highway. This requires special treatments within and adjacent to the barrier. Options for addressing drainage issues in the vicinity of noise barriers are discussed below.

7.1.1 Use of Barrier Overlap Section to Accommodate Drainage Flows.

Where acoustical requirements permit it and geometrics allow it, a barrier overlap section can be constructed. While such sections are normally built to provide access gaps in longer sections of noise barriers, they can be well suited as a means of carrying water past and through a barrier (see Figures 185 and 186). Additional discussion related to barrier overlap is included in Section 9.4.1. Details related to acoustical requirements of such overlaps are discussed in Section 3.5.6.1.

Figure 185. Drainage: use of barrier overlap (data base #901).

Figure 186. Drainage: use of barrier overlap (data base #382).

7.1.2 Accommodating Water Flow Through a Barrier.

A variety of sizes and shapes of openings in barriers have been employed to carry water (which would otherwise pond on one side of the barrier) through the barrier (see Figures 187 and 188). The essential consideration in this type of design is to ensure that the size and location of such openings do not result in any significant degradation of the barrier's acoustical performance. The effect of a continuous gap of up to 20 cm (7.8 in) at the base of a noise barrier is usually within 1 dB(A).[32] An important consideration is to ensure that proper protection in the form of grates or bars is provided to restrict entry by small animals (cats, small dogs, etc).

Figure 187. Drainage: water through a barrier (data base #1414).

Figure 188. Drainage: water through a barrier (data base #1132).

7.1.3 Accommodating Water Flow Along and/or Underneath a Barrier.
Several techniques have been used to carry water underneath a barrier. The most common is via the use of a swale system running parallel to the barrier (see Figure 189 and 190), or feeding water to catch basins (inlets) through which pipes running underneath the barrier are connected (see Figures 191 and 192). Depending upon the grades in the vicinity of the barrier, such swales may be required on either the highway side or the adjacent property side of the noise barrier. These pipes may carry water to a discharge point onto the ground surface or to the closed pipe or open **culvert** drainage system of the highway.

Figure 189. Drainage: water along a barrier (data base #984).

Figure 190. Drainage: water along a barrier (data base #1065).

Figure 191. Drainage: water underneath a barrier (data base #1798).

Figure 192. Drainage: water underneath a barrier (data base #1799).

Another method of carrying water underneath a barrier is by providing a porous stone trench beneath the base of the barrier through which water can seep in a manner similar to that of a **subdrain**. An adaptation of this treatment involves erecting the barrier with its panel bottoms raised several inches above the adjacent ground elevation and then mounding porous stone on both sides of the barrier to close the gap while also allowing water to pass through (see Figure 193). These porous stone systems require maintenance to ensure they do not seal up over time with sediments. These types of treatments must be compatible with the type of noise barrier erected. A barrier erected with a continuous footing will preclude the use of either of the porous stone techniques.

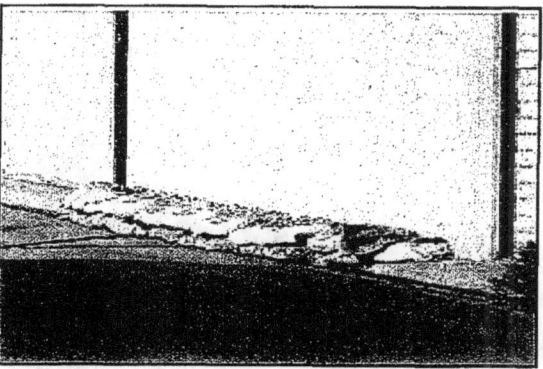

Figure 193. Drainage: water underneath a barrier (data base #2071).

7.1.4 Special Drainage Considerations in Flood Plain Areas.

In some instances, it is necessary to construct noise barriers in areas subject to flooding conditions in which flood water would actually flow across the highway. Since a noise barrier would naturally restrict such flow and exacerbate the water flooding condition, the bottom panels of a precast concrete panel system can be hinged so that the pressure of any built-up water would swing the panels outward, allowing the release of flood waters (see Figure 194). Following the passage of the flood water, the panels would swing back into a vertical position.

Figure 194. Drainage: water through a swing panel (data base #6168).

7.2 Lighting, Sign Supports and Utility Poles and Other Elements Mounted on or Adjacent to Noise Barriers

Where noise barriers are required to be constructed in the vicinity of such features, a variety of techniques have been employed (see Figure 195). In some cases, concrete cast-in-place noise barriers have been poured right up to concrete utility poles. In other cases, walls have been offset or have had inserts constructed to accommodate utility elements such as transformers and poles. While major sign supports are accommodated in a similar manner, supports for smaller signs (such as yield signs) as well as other features such as call boxes have been mounted directly to the facade of the noise barrier walls (see Figures 196

Figure 195. Lighting incorporated into a barrier system (data base #1180).

and 197). When mounting such elements on noise barriers, care needs to be taken to ensure that adequate horizontal and vertical clearances are provided.

Figure 196. Other elements mounted on a barrier (data base #1074).

Figure 197. Other elements mounted on a barrier (data base #1432).

7.3 Effects of Underground Utilities on Noise Barrier Design and Location

The presence of underground utilities can have a major bearing on the type of noise barrier designed and the location of such a barrier. Significant utilities in the path of a barrier can preclude the use of certain types of deep footings (e.g., **pile** type footings) and require use of either a shallow spread footing or a barrier design which requires no footing (see Figure 198). An undulating noise wall configuration (in plan view) may be advantageous in "jogging around" underground utilities (see Figure 199).

Figure 198. Effects of underground utilities on barrier design (data base #1089).

Figure 199. Effects of underground utilities on barrier design (data base #2228).

7.4 Effects of Overhead Utilities on Noise Barrier Design and Location

The presence of overhead utilities may restrict the size of components used in construction of a barrier (see Figures 200 and 201). If vertical clearances are limited, smaller precast elements (such as stacked panels) and smaller lifting equipment may be required. In certain instances, the use of a crane or any heavy construction equipment may be precluded altogether by vertical clearance restrictions. In such cases, precast panel designs may be excluded as an option or may require special posts which enable the panels to be placed from the side rather than set in from above (see Figure 202). Barrier type may also be limited to a block wall, a cast-in-place wall, or some other system which can be constructed without the use of lifting or heavy construction equipment. Consideration of both the post and panel elements must be made.

Figure 200. Effects of overhead utilities on barrier design (data base #835).

Figure 201. Effects of overhead utilities on barrier design (data base #5288).

Figure 202. Effects of overhead utilities on barrier design (data base #869).

Section Summary

Item #	Main Topic	Sub-Topic	Consideration	See Also Section	✓
7-1	Drainage	Barrier Overlap	Ensure the acoustical requirements of such overlaps are met.	3.5.6.1 7.1.1	
		Accommodating Water Through a Barrier	Ensure that the size and location of openings do not result in degradation of acoustical performance, and also ensure that protection in the form of grates or bars is provided to restrict entry by small animals.	7.1.2	
		Accommodating Water Under a Barrier	If a porous stone systems is used, maintenance is required to ensure it does not seal up over time with sediments.	7.1.3	
		Flood Plain Areas	Include possible design additions where the bottom panels of a precast concrete panel system can be hinged so that the pressure of any built-up water would swing the panels outward, allowing the release of flood waters.	7.1.4	
7-2	Lighting, Sign Supports, Utility Poles, Etc.		When mounting traffic or safety-related elements on or adjacent to noise barriers, care needs to be taken to ensure that adequate horizontal and vertical clearances are provided.	7.2	
7-3	Underground Utilities		Utilities in the path of a barrier can preclude the use of certain types of deep footings and require use of either a shallow spread footing or a barrier design which requires no footing. An undulating noise wall configuration may be used to "jog around" underground utilities.	7.3	
7-4	Overhead Utilities		If vertical clearances are limited, stacked panels and smaller lifting equipment may be required - or the barrier type may be limited to a block wall or a cast-in-place wall.	7.4	

8. STRUCTURAL CONSIDERATIONS

Proper design of noise barrier systems requires the consideration of a variety of structurally related factors. This section is not intended to provide either a standard or a recommended process for the structural design of noise barriers. Rather, its goal is to identify structural issues which should be addressed and considered in the design of barrier systems. Specific application and interpretation of appropriate structural criteria is the responsibility of the respective responsible organization in charge of designing and constructing the barrier system.

8.1 Expansion and Contraction of Barrier Materials

All materials used in the construction of noise barriers expand and contract with temperature and moisture variation. Such expansion and contraction must be appropriately considered in the design of all elements of noise barrier systems. Failure to do so can result in both structural, acoustical, and aesthetic problems. The individual barrier elements themselves must be designed and constructed to preclude unacceptable deformation, cracking etc. Conditions where consideration of such expansion and contraction effects is most essential include:

- **Post-to-panel connections** - Expansion/contraction is normally accommodated by allowing sufficient space or gaps between the post web and the panel. Some designs may call for full or partial caulking or shimming of the panel/post flange contact point to ensure adequate load transfer and/or to avoid sound leakage. Care must be taken to ensure that the caulking and/or shimming material does not restrict panel expansion or contraction (see Figure 203).

Figure 203. Expansion and contraction of materials: post to panel connections (data base #8038).

- **Panel-to-panel connections** - Such connections occur both horizontally (such as between stacked panels or tongue and grove panels) and vertically (such as with post-less or vertical tongue and grove barrier systems). Such connections must allow sufficient movement while maintaining tight joints (see Figure 204).

- **Expansion joints on cast-in-place and brick/masonry noise barrier systems** - Vertical expansion joints are required at sufficient intervals to preclude cracking of the wall system. Designing of such joints in a manner which ensures aesthetic and acoustical integrity is often a challenge.

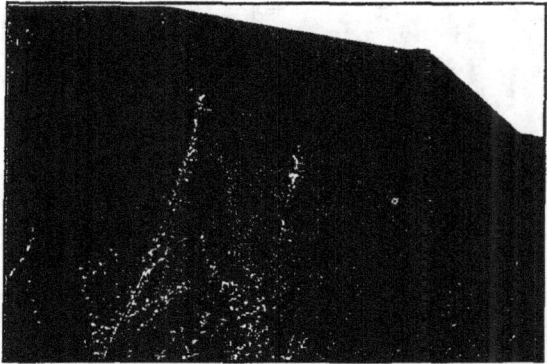

Figure 204. Expansion and contraction of materials: panel to panel connections (data base #181).

- **Connections between ground-mounted and structure-mounted barriers** - It is sometimes necessary for a ground-mounted barrier to continue onto a structure (bridge or retaining wall). In such instances, special detail barrier sections are required in order to accomplish an adequate connection which is both structurally and acoustically sound while maintaining the desired barrier aesthetics (see Figures 205 to 207).

Figure 205. Expansion and contraction of barrier materials: connections between barriers (data base #463).

Figure 206. Expansion and contraction of barrier materials: connections between barriers (data base #1716).

Figure 207. Expansion and contraction of barrier materials: connections between barriers (data base #1720).

- **Structure-mounted barriers** - In addition to the expansion considerations discussed above, barriers mounted on structures must also accommodate expansion/contraction at the structure's expansion joint locations (see Figures 208 and 209).

Figure 208. Expansion and contractions of barrier materials: structure barriers (data base #414).

Figure 209. Expansion and contraction of barrier materials: structure barriers (data base #1711).

8.2 Noise Barrier Loadings

Design of noise barrier systems must include consideration of a variety of design loads, both individually and in combination with each other. Such loads include:

- **Dead Load** - The weight of the barrier itself must be considered in all barrier design calculations. Weight considerations are particularly critical in the design of structure-mounted barriers and can require modifications to the structure design itself. Lightweight barrier materials are often utilized in situations where existing or proposed structures are limited in the amount of additional weight which they can accommodate. Ice loads represent a special type of dead load caused by water freezing and building up on exposed barrier surfaces.

- **Wind Load** - Wind loads vary with geographic location and can be influenced by elevation in relation to existing topography. They affect the overturning moment or rotational force placed upon the barrier, its foundation, and/or the structure to which the barrier is attached. Unlike dead loads, wind loads are essentially the same, regardless of barrier material type.

- **Snow Loads** - Unlike ice loads, snow loads are not considered to be dead loads placed upon the barrier. Rather, in barrier design, considerations related to snow relate to the generally horizontal forces of both plowed and stored snow which can be placed on the vertical surface of the barrier. Design of the barrier to accommodate such loadings should consider the area available for safe storage of plowed snow as well as the relationship (both horizontally and vertically) of the barrier to the location of snow clearing and snow removal equipment (plows, front end loaders, melters, and blowers).

- **Impact Loads** - Impact loads can be classified as loads placed on the barrier due to errant vehicles and airborne debris. Noise barriers are not themselves designed to withstand the full force of a vehicle impact. Rather, either a protective metal guardrail or a Jersey-type barrier placed in front of the noise barrier is usually relied upon to keep errant vehicles away from the barrier. Placement of a noise barrier on a structure is usually restricted to the structure's parapet. In such cases, options for barrier mounting to the parapet (either top or face mounting) should be reviewed in light of the potential for the barrier being impacted by all portions of vehicles, including the potential impact resulting from a tall truck tilting toward the barrier after hitting the protective barrier. Airborne debris such as retreads, stones, vehicle parts, etc., can also strike the barrier, regardless of what type of protection is provided against vehicle impact. The impact of such strikes upon the barrier is mainly a function of the durability of the noise barrier's material, especially its surface. While lightweight materials have an obvious positive influence in terms of dead load design factors, they may not be as durable in terms of impact as compared to heavier barrier materials.

8.3 Barrier Height Considerations

Barrier heights can be influenced by one or all of the above load-related factors and by other conditions such as the presence of overhead utilities or other restrictions, cost, aesthetics, and foundation requirements.

8.4 Foundation Requirements

Footings and foundations for ground-mounted noise barriers are typically limited to concrete cylinders (caissons), spread footings, and continuous footings. When designing these, the following factors contribute to the selection of the type of footing to be used as well as its depth and size:

- The bearing capacity and compressibility characteristics of the surrounding soil or rock;
- Possible ground movements;
- Anticipated future excavation activity adjacent to the foundations;
- Ground water levels;
- Extent of frost penetration;
- Extent of seasonal volume changes of cohesive soils;
- The proximity and depth of foundations of adjacent structures; and
- Overall ground stability, particularly adjacent to cut or fill slopes.

8.4.1 Concrete Footings in Earth.

Concrete for drilled or augured footings should be cast entirely against undisturbed soil. If other than drilled footings are necessary, the footings should be formed and the excavation should be backfilled with granular materials and properly compacted. The tops of all footings should be shaped to provide for full horizontal seating of panels with the remaining surface area to be sloped away from the post so as to shed water. Stepped footings may be required to suit grade changes (see Figure 210). To avoid premature failure of the concrete in the footings, the concrete should have an opportunity to cure properly before the noise panels are installed.

Figure 210. Stepped concrete footings in earth (data base #2939).

8.4.2 Concrete Footings in Rock.

When rock is encountered, a different technique should be considered to ensure a stable foundation. A typical example would be to construct the footing in the same manner as for footings in earth with partial embedment into solid rock. All excavations into rock should be backfilled with either concrete or other suitable material. The excavation above the top of rock may be backfilled with granular material.

Section Summary

Item #	Main Topic	Sub-Topic	Consideration	See Also Section	
8-1	Expansion and Contraction of Materials	Post-to-Panel Connections	Care must be taken to ensure that caulking and/or shimming material do not restrict panel expansion or contraction.	8.1	
		Panel-to-Panel Connections	Care must be taken to allow sufficient movement within panel-to-panel connections while maintaining tight joints.	8.1	
		Expansion Joints	Vertical expansion joints are required at sufficient intervals to preclude cracking in cast-in-place and brick/masonry barrier systems.	8.1	
		Ground-to-Structure-Mounted Connections	Consideration must be given to connections between ground-mounted and structure-mounted barriers.	8.1	
		Structure-Mounted	For structure-mounted barriers, consideration must be given to the expansion/contraction at the structure's expansion joint locations.	8.1	
8-2	Noise Barrier Loadings	Dead Load	Consider possible modifications to the structure design to accommodate barrier weight.	8.2	
			Consideration must be given to ice loads caused by water freezing and building up on exposed barrier surfaces.	8.2	
		Snow Load	Consider the need for area available for safe storage of plowed snow and the location of the barrier for snow clearing and removal equipment.	8.2	
8-3	Barrier Height Limitations	Aesthetics	Because of the potential closeness of barriers, reduce the visual dominance of a very tall barrier by locating the barrier at least 2-4 times its height from the nearest receiver.	6.1.7 8.3	
		Drainage and Utility	Barrier height may be limited by the presence of overhead utilities.	7.4 8.3	
		Structural	Consideration must be given to foundation requirements of tall barriers.	8.3	
8-4	Foundation Requirements	Earth vs. Rock	The following factors contribute to the selection of the type of footing to be used as well as their depth and size: ■ The characteristics of the surrounding soil or rock; ■ Possible ground movements; ■ Anticipated future excavation activity adjacent to the foundations; ■ Ground water levels; ■ Extent of frost penetration; ■ Extent of seasonal volume changes of cohesive soils; ■ The proximity and depth of foundations of adjacent structures; and ■ Overall ground stability, particularly adjacent to cut or fill slopes.	8.4	

Item #	Main Topic	Sub-Topic	Consideration	See Also Section	✓
		Concrete Footings in Earth	Concrete for drilled or augured footings should be cast entirely against undisturbed soil. The concrete should have an opportunity to cure properly before the noise panels are installed.	8.4.1	
		Concrete Footings in Rock	All excavations into rock should be backfilled entirely with concrete.	8.4.2	

9. SAFETY CONSIDERATIONS

This section describes the various safety considerations related to noise barrier design. Safety is a factor which must be given appropriate consideration in the design of any noise barrier system. While organization and/or industry standards are adhered to in the design of noise barriers, these standards are usually geared to providing acceptable designs related to factors such as dead loads, overturning loads, durability, strength of materials, etc.

9.1 Qualitative Evaluation of Safety

Barriers should be located (where possible) so as to be somewhat protected from vehicular impact, especially since they are not specifically designed to withstand severe collisions. In addition, vehicles should be provided the appropriate protection against impacting a barrier. Since no known standards or formulas exist to specifically determine "how safe is safe" in terms of noise barrier design associated with impacts, decisions related to determining the "appropriate" degree of protection to provide in the construction of barriers in any given situation may be based on the "best engineering judgment" of the designer. Unfortunately, but understandably, such decisions are sometimes influenced by (and in some cases totally based on) individual opinions and concerns, but with little or no quantitative or qualitative assessment of risks, probability of impact occurrence, or other evaluations.

Although a topic requiring research attention, the development of a quantitative formula or standard related to the above discussed issues is beyond the scope of this Handbook. However, a qualitative evaluation with the consideration of the following factors is considered imperative in the design of noise barriers where safety concerns exist.

9.1.1 Evaluate the Need for Special Considerations Related to Safety.
The first step in the qualitative evaluation process should involve determining whether or not any special modifications to the normal noise barrier design are warranted. Things to consider in this determination include, but are not limited to:

9.1.1.1 Probability of Occurrence of the Barrier Being Impacted.
In this context, "impact" should be considered a force sufficient to result in the barrier either being knocked down or shattered to the point where the resultant debris becomes a safety concern. One should investigate the history of barriers being impacted in similar orientations elsewhere (nationally and/or statewide) and evaluate, based on the known accident history of the particular highway (if barrier is to be placed on an existing highway) or similar area highways (similar volumes, composition, configuration, etc.), the probability of impact.

9.1.1.2 Consequences of the Barrier Being Impacted.
Assuming that the evaluation in Section 9.1.1.1 concludes that a high enough probability of occurrence exists (where "high enough" is defined by the responsible organization) to warrant further special considerations, an evaluation of the consequences of barrier impacts should be undertaken. In areas where a ground-mounted noise barrier is proposed, the area immediately behind the noise barrier should be evaluated. If this area happens to be

an actively used portion of a backyard, a school playground, or frequented area, the consequences of impact will be greater than in portions of such properties with little or no regular activity (see Figure 211). In areas where a noise barrier is proposed to be erected on a bridge or viaduct section, the frequency and type of use beneath and adjacent to the bridge should be identified. Fewer consequences are likely in areas where the barrier passes over or adjacent to low volume roadways or relatively inactive lands as compared to high volume roadways and active land uses. Consideration should also be given in this evaluation to the added protection provided by the barrier in terms of retaining objects (vehicle parts, vehicle cargo, stones, road salts, etc.) which, in the absence of the barrier, would pose a potential hazard to land uses adjacent to or beneath the highway. While this discussion has focused on the consequences to adjacent land users, the consequences of barrier impact need to also be evaluated in terms of the drivers and passengers of vehicles operating on the highway.

Figure 211. Consequences of a barrier being impacted (data base #1148).

9.1.2 Modifications to the Noise Barrier Design.

Assuming that the evaluation in Section 9.1.1.1 indicates that sufficient consequences exist to warrant further special considerations, modifications to the standard noise barrier design should be evaluated. Such modifications may relate to the barrier's location, attachment/reinforcement details, type, and protective devices as discussed below. As each modification is evaluated, a re-evaluation of the applicable factors in Section 9.1.1 should be performed.

- **Barrier location** - The location of the barrier may be able to be modified to place it in a position where it is less vulnerable to impact by vehicles. If conditions (topography, acoustic requirements, drainage, etc.) permit, a ground-mounted barrier at the property line of an active land-use area may be moved closer to the highway. If structural conditions permit, a noise barrier designed to sit on top of the bridge parapet may be repositioned to be mounted on the outside face of the parapet, giving added protection from potential impacts of automobiles and trucks.

- **Barrier attachment/reinforcement details** - Barrier component attachment and/or reinforcement details may be modified. As shown in Figure 212, cables can be placed through pre-drilled/pre-formed holes in posts and either run through horizontal conduits in concrete and composite panels or attached to panels in metal and wood systems. These cables thus tie a barrier system together and provide a means of retaining pieces of a barrier damaged by a vehicular impact. Additional

Figure 212. Barrier attachment/reinforcement details (data base #1267).

reinforcement rods or mesh can be added to concrete, masonry block and brick noise barriers to increase the panels' strength and reduce the size of shattered pieces. Similarly, additional framing can be added to wood, metal, and composite barrier systems. Such additions add weight to the barrier system, a factor which could become a problem on structure-mounted barriers. In addition, they could also compromise barrier aesthetics.

- **Barrier type** - The type of barrier may be modified in terms of its material type and/or configuration. Certain barriers can be designed without posts or with concealed posts. These "postless" systems reduce the potential for errant vehicles to "snag" a protruding post. In many cases, lighter-weight materials have been selected for barriers located in areas deemed to be of higher risk to adjacent land users. This is particularly true on structures, where such lighter-weight systems have an obvious advantage in terms of structure load designs. The somewhat common conception that lighter-weight barriers will be less likely to cause damage if knocked down or if knocked off a bridge requires a more rigorous evaluation. Such systems can still require quite heavy metal, wood, or concrete posts as well as have components which, if dislodged, can tear and "sail" larger distances than heavier panel components. Lighter-weight barrier systems tend also to be more susceptible to damage from more frequent, albeit less major, impacts from vehicular scrapes, vehicle parts, vehicle cargo, stones, road salts, etc. The potential exists for such lighter-weight panels to be knocked down by a collision which could have been otherwise withstood by a heavier, more substantial barrier. Obviously, the maintenance and durability implications of any barrier modification need to be thoroughly evaluated in the decision-making process.

- **Barrier protective devices** - In addition to the considerations listed above, the potential for a noise barrier being impacted can be reduced by the placement of a protective barrier (steel guard rail, concrete Jersey barrier, etc.) between the noise barrier and the highway traffic or by erecting a higher than normal safety shape barrier in front of a noise barrier in close proximity to traffic. Since protective barriers are themselves considered an obstruction, any such protective barrier must be designed in compliance with appropriate standards. In addition, the consequences of impact of a vehicle hitting the protective barrier must be weighed against impact consequences which would exist in the absence of such a protective device. Section 9.3 provides a more detailed discussion on protecting a barrier from traffic.

9.1.3 Overall Results of Qualitative Evaluation.

By objectively considering the above factors, a rational decision regarding the appropriate consideration of safety issues and concerns can be made consistent with the goals and criteria related to acoustical, structural, and aesthetic aspects of noise barrier design. Through such an evaluation, the appropriate justification can be documented and the appropriate safety-related noise barrier modifications can be implemented on a case-by-case basis.

9.2 Sight Distance

Sight distance (as impacted by noise barriers) is a factor requiring consideration along horizontal curve sections of highways and at locations where a barrier terminates near a highway's or ramp's intersection with another roadway. Noise barriers (either wall or berm) placed in either the median or along the outside of a highway can impact sight distance. Solutions to providing adequate sight distance include additional setback of the barrier, curving the barrier back approaching intersection (see Figure 213), or terminating the barrier at a point short of its required acoustical end point.

Figure 213. Sight distance (data base #612).

9.3 Traffic Protection

Noise barriers are protected from vehicular impacts when constructed within the recovery zone (clear zone) normally provided for vehicles. Devices used to provide such protection include metal or wood guard rails (or guide rails) or concrete safety-shaped (Jersey barriers) protective barriers. Metal and wood guardrails are placed at a distance in front of the noise wall equal to or greater than the maximum deflection of the guard rail (see Figure 214). Concrete Jersey barriers are sometimes placed immediately in front of a noise barrier, but are often placed some distance in front of the noise wall with the area between the Jersey barrier and the noise barrier sometimes filled with stone or earth. This latter treatment provides a greater degree of protection and also allows for plantings in the area between the Jersey barrier and the noise barrier (see Figure 215). Certain cast-in-place concrete and post-and-panel noise wall systems incorporate the Jersey barrier shape in the bottom portion of the noise wall as either an integral portion of the noise wall or as the bottom panel of the noise wall (see Figures 216 and 217). Where the noise wall is potentially subject to vehicular impact on both sides (such as where a parallel frontage road or a local perpendicular street exists) such protection may be required on both sides of the noise barrier.

Figure 214. Traffic protection (data base #1251).

Figure 215. Traffic protection (data base #64).

Figure 216. Traffic protection (data base #4).

Figure 217. Traffic protection (data base #8036).

9.4 Emergency Access

Noise barriers often interrupt the path between the highway and adjacent local roadways. During emergencies (accidents, spills, fires, etc.) access from these local roadways is often necessary and/or desirable, particularly on stretches of limited access highways with significant distances between interchanges. Barrier access points for emergency or maintenance situations are typically identified with signs on both sides of the barrier to enable the proper coordination of emergency/maintenance personnel. Such signs may be located either on the barrier itself and/or along the highway and the adjacent roadway.

9.4.1 Barrier Overlap Sections. Access ways can be provided by overlapping barriers (see Figures 218 and 219). The distance between the walls of the overlapping section (measured perpendicular to the barriers) is typically dictated by the size of equipment which may need to pass through the access opening; whereas, the length of the overlap is almost always a function of acoustics such that the overlap length is designed to maintain the amount of noise leak at an absolute minimum. Section 3.5.6.1 of this manual discusses the acoustical concerns related to barrier overlap sections. Another consideration related to the design of barrier overlap sections is the potential for increased crime in the immediate areas surrounding the overlapping sections, particularly where a pedestrian overpass is also located nearby. To address this concern, safety measures, including additional lighting or a modified overlap design to provide more open visibility, may need to be implemented.

Figure 218. Barrier overlap sections (data base #109).

Figure 219. Barrier overlap sections (data base #382).

9.4.2 Access Doors. Access needs can also be met by providing doors in noise barriers at appropriate intervals. These doors need to be designed so as to be acoustically "sealed" when closed. When not in use, doors are almost always locked to avoid unauthorized access. Several techniques are used to enable authorized users to gain access through doors. They vary from combination locks to common (master) key systems. Materials used to make doors are usually either metal or wood. As such, the door's material may be different from the material of the overall barrier. This fact suggests that the noise wall designer should pay particular attention, during the design stage, to aesthetics of the access doors in relationship the remainder of the barrier (see Figures 220 to 225).

Figure 220. Access door (data base #1519).

Figure 221. Access door (data base #1631).

Figure 222. Access door (data base #6519).

Figure 223. Access door (data base #1539).

Figure 224. Access door (data base #2344).

Figure 225. Access door (data base #2414).

On rare occasions, it may be necessary to design a noise wall section or sections to be temporarily removed. An example of such a situation could be related to a noise wall constructed next to a utility sub-station where the only access for large equipment is from the highway along which the noise wall runs. Noise walls can be constructed with their panel lifting inserts left intact and with sufficient space between posts to accommodate the necessary passage of equipment, repair parts, etc (see Figures 226 and 227).

Figure 226. Emergency access opening (data base #5017).

Figure 227. Emergency access opening (data base #2290).

9.5 Fire Safety

Noise barriers may also interrupt the path between the highway and a source of water required to be accessed in the event of a fire or spill on the highway. Such water sources may be a pond, lake, stream, or fire hydrant. Since fire hoses cannot be practically draped over a noise barrier, special design considerations are required. Where neither barrier overlaps nor access doors (see Figure 228) are available for running fire hoses, emergency access openings or valves can be incorporated directly into the design of the noise wall panels.

Figure 228. Access for fire (data base #1416).

The following is a brief description of some of the more successful designs used to date.

- **Hose Couplers Incorporated into Noise Barrier Panels** - This technique involves incorporating one or more couplers directly into the noise barrier panel (see Figures 229 and 230). This fixed coupler allows the connection of fire hoses on both sides of the noise barrier, and effectively eliminates any kinks in the hose.

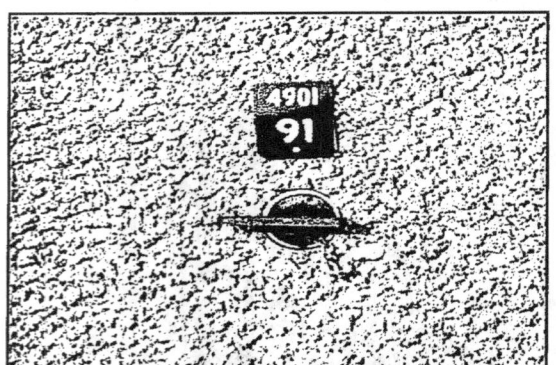

Figure 229. Access for fire (data base #54).

Figure 230. Access for fire (data base #260).

A critical consideration with this type of a treatment is to have the correct size connection (diameter, thread size, etc.) for all fire companies which may need to access the connection. Also, strength of the panel, the adjacent posts, and, in particular, the post-to-panel connections must be analyzed to ensure that they are capable of withstanding the thrust loads generated by the force of the moving water in the hose lines.

- **Panel-Mounted Valves** - Another technique is to include a suitable size valve into the panels when they are being fabricated (see Figure 231). This method should also be structurally evaluated to ensure the capability of withstanding thrust loads. Another concern that should be considered with this type of system is that the closure caps (required to keep debris and other objects out of the connections and to protect the coupling threads) and their retaining chains, being brass, can become the target of theft and vandalism. Where elevation differences occur, the use of dry stand-pipes may be required (see Figure 232)

Figure 231. Access for fire (data base #8089).

Figure 232. Access for fire (data base #1714).

- **Small Covered Openings** - Another treatment involves casting an opening of sufficient size (typically about one foot square) to allow the pass through of several fire hoses. Such a treatment is compatible with any size hose and also allows for easy communications between emergency personnel on either side of the barrier. Passage of small tools (wrenches, axes, etc.) from one side of the barrier to the other is also enabled. This type of treatment requires some form of device (door, flap, etc.) to seal the opening when it is not in use in order to avoid acoustic degradation and to restrict the passage of small animals, etc. While early designs of such closure devices experienced vandalism and theft problems, recent designs have proven to be adequate. They include flaps, knock-out sections cast directly into the panels, and closures requiring special tools to open (see Figures 233 and 234).

Figure 233. Access for fire (data base #1568).

Figure 234. Access for fire (data base #1112).

Whichever system is used, identification signs are required on both sides of the barrier to enable the proper coordination of emergency personnel. Such signs may be located either on the barrier itself and/or along the highway and the adjacent roadway. The choice of an opening in the wall versus a valve through the wall is usually a determination made with significant input from the local fire departments.

9.6 Glare

Glare is generally a problem on noise barriers with smooth surfaces, such as metal and transparent barriers (see Figure 235). It is more prevalent on lighter colored surfaces and can be a problem in daytime (low sun angle) and nighttime (due to headlights) periods and is particularly bad during nighttime periods when the barrier may be wet. Use of rougher types of surface treatments and deep relief patterns can reduce or eliminate glare impacts.

Transparent noise barrier panels, or any other high gloss finish on panels, such as plastics and metals, reflect light. In unfavorable situations, by day or night, the reflection of light from these glossy surfaces can be

Figure 235. Glare (data base #392).

troublesome and even dangerous for drivers, either because of glare or simply by the reflection of another vehicle's image on the glossy surface. These problems arise, in particular, in the case of curved roadways or where the traffic passes close to the wall, resulting in a low-angle incidence of light.

Consequently, in constructing high-gloss surfaced noise barriers, every precaution must be taken to ensure that such reflections do not reduce the safety of the roadway.

- **Sources of Reflections:**

    By Day - During the day, the reflections are produced typically by the sun, and its effects are most harmful when it is low on the horizon, namely in the morning and in the evening. The sun's rays, bounced back onto traffic, may momentarily blind a driver. This phenomenon, though of short duration, is bound to occur and its intensity is typically such that it cannot be neglected.

    By Night - The major source of reflections during night-time is normally from vehicle headlights, whose rays generally strike the reflective panels at a low angle of incidence. In an urban environment, the barriers are often situated within infrastructures where the roadway is illuminated and traffic flow is controlled by lighted signs and traffic signals. Reflections could also be coming from adjacent roadways as well.

    These reflections are directly affected by the angle of incidence. If the angle is less than 40 degrees, approximately 8 percent of the intensity of the incident beam is reflected. This is generally the case for vehicles close to the noise barrier. If the angle of incidence is greater than 40 degrees, the reflection has virtually the same intensity as the incident beam. This occurs more on curved roadways.

To overcome some of the light reflection problems, the following solutions should be considered:

- **Tilted Walls** - By tilting the entire wall between 6 and 20 degrees, the reflected rays can be deflected away from the oncoming traffic, thus removing the potential hazard caused by the glare. The most effective angle must be determined on a site-by-site basis.

- **Tilted Stacked Panels** - An alternative is to construct the wall system using a stacked panel arrangement and only tilting the individual panels. This will provide the same results as tilting the entire wall. It may also be possible to only tilt specific portions of the wall. Tilted stacked panels are also referred to as "clapboard."

- **Protrusions** - This solution requires that the panels be mounted on the posts in such a way that either the posts or a similar deep protrusion will, at some distance, hide the surface of the panels from the traffic.

- **Anti-glare Screens** - Another option to constructing a tilted noise barrier is to install anti-glare screens in the median to remove the problem of glare from the oncoming traffic. However, dependent on the type of material used, the screen may alter the acoustical characteristics of the roadway. Therefore, the desirable type of material to use in this situation would be an "expanded web" type made of either metal or plastic which would not interfere with the acoustics. This type of anti-glare screen material could also be installed adjacent to the transparent panels. Although this would provide shielding from reflected light, the expanded metal is always visible to some degree, regardless of the angle it is being viewed through, and, thus, may reduce some of the benefits of a transparent barrier unobtrusiveness for which the barrier was designed.

9.7 Shatter Resistance

When a noise barrier is impacted by a vehicle, the effect of barrier components shattering is a concern. The effect is of particular concern where the barrier is located on a structure overpassing another highway causing potential injury to nearby vehicles, pedestrians, or adjacent residents. For each of these situations, the shatter resistance of the noise barrier components should be evaluated in conjunction with the probability of vehicular impact. Differing opinions exist on allowing the components to shatter into small pieces with little or no potential for causing injury, as opposed to totally preventing the components from shattering by installing additional reinforcements.

9.8 Icing and Snow Removal

Design of noise barriers in climates subject to ice and snow conditions should consider the placement of barriers a sufficient distance from the travel way to ensure adequate space for storage of plowed snow and to ensure that the barrier can withstand the additional loads that may result from plowed snow being both thrown and piled up against the barrier. Barriers also will shade portions of highways at some time during the day. Depending upon the particular orientation of the barrier and its height, this shading can result in significant interference of the normal sunlight melting of ice or snow on the highway's shoulders and travel lanes. These factors should be considered in the design process so that, in critical areas, the possibilities of wide shoulders and minimal barrier heights can be incorporated.

Section Summary

Item #	Main Topic	Sub-Topic	Consideration	See Also Section
9-1	Qualitative Evaluation of Safety	Modifications to Barrier Design	Consider the safety implications of a barrier's location.	9.1.2
			Reinforcement additions add weight to the barrier which can be a problem on structure-mounted barriers.	9.1.2
			Consider the safety implications of a barrier's type.	9.1.2
			Consider barrier protective devices as part of the overall barrier design.	9.1.2
9-2	Sight Distance		Sight distance concerns along horizontal curve sections of highways and at locations where a barrier terminates near a highway's or ramp's intersection with another roadway.	9.2
9-3	Emergency Access	Barrier Overlap	Ensure the acoustical requirements of such overlaps are met (see Table Item 3-2).	3.5.6.1 9.4.1
			Ensure the safety of pedestrians within or near the gap.	9.4.1
		Access Doors	Doors need to be designed so as to be acoustically "sealed" when closed. When not in use, doors should be locked to avoid unauthorized access.	9.4.2
		Emergency Access for Utilities	Careful consideration must be given to the design of large utility access doors, which over time, have a tendency to warp due to their massive size.	9.4.3
9-4	Fire Safety	Hose Couplers	Ensure the correct size connection and ensure the capability of withstanding the thrust loads generated by the force of the moving water in the hose lines.	9.5
		Panel-Mounted Valves	Ensure the capability of withstanding thrust loads, and keep in mind that the closure caps and their retaining chains can become the target of theft and vandalism.	9.5
		Small Covered Openings	A device is needed to seal the opening when it is not in use in order to avoid acoustic degradation and to restrict the passage of small animals.	9.5
			Note that identification signs are required on both sides of the barrier to enable the proper coordination of emergency personnel.	9.5
9-5	Glare	Tilted Walls	Walls tilted for acoustical reasons may cause glare.	4.1.2.1.1 9.6
		Anti-glare Screens	Dependent on the type of material used, the screen may alter the acoustical characteristics of the roadway.	9.6
9-6	Shatter Resistance	Barriers on Structure	Particular concern should be given for barriers on structure overpassing another highway which can cause potential injury to nearby vehicles and pedestrians.	9.7
		Safety	Consideration should be given to the potential shatter resistance of transparent and plastic barrier materials.	4.2.1, 5.5 5.6, 9.7
9-7	Icing and Snow Removal	Structural	Ensure that the barrier can withstand the additional loads that may result from plowed snow being both thrown and piled up against the barrier.	8.2 9.8

FHWA Highway Noise Barrier Design Handbook — *Safety Considerations*

Item #	Main Topic	Sub-Topic	Consideration	See Also Section	✓
		Maintenance	Place barriers a sufficient distance from the travel way to ensure adequate space for storage of plowed snow.	9.8	

10. PRODUCT EVALUATION

This section deals with the process of evaluating noise barrier products designed and manufactured privately before they are installed. The evaluation of new products for possible use in highway construction is not a new concept for highway and transportation agencies. In fact, the evaluation of new materials and processes was often a principal activity in the highway agencies that were formed at the beginning of highway expansion following the introduction of the automobile. As practice became more standardized, principal attention shifted to other matters, and new product evaluation became a routine function to be initiated and undertaken along with other activities by staff members of various transportation agencies.

New product evaluation offers the best opportunities for control and standardization of the evaluation process. It also offers the opportunity to apply budgeting and funding practices that permit the measurement of productivity and benefits accruing from new product evaluation.

Most agencies have adopted a one-window concept for this type of external contact. That is, all evaluations are carried through with the overall responsibility assigned to a single division head regardless of where the evaluation takes place. This concept also ensures the focalization of all vendor activities to a single area, which reduces the number of unnecessary calls by vendors to several divisions and offers the opportunity for improving control over new product evaluations.

10.1 Evaluation Process
The total process of evaluating new noise barrier products can be broken down into the following six steps:

10.1.1 Step 1 - Submission.
The presentation of the noise barrier design submission for acceptance to a transportation agency is probably the most critical stage in getting a noise barrier product approved. It can also be the most frustrating for both the manufacturers and the approving agencies. The key is knowing what the requirements and specifications are for acceptance by the responsible organization. And, just as important, in what format are the agencies accustomed to seeing these submissions?

Since transportation agencies are dealing with public funds, they are obligated to ensure that every product used on a roadway meets durability, safety, and functionality requirements. In addition, they should address life cycle costing and be cost effective. The methodologies used to meet these requirements vary substantially between agencies. Many agencies have adopted, to some degree, the approval strategy and principles outlined in the National Cooperative Highway Research Program, Synthesis of Highway Practice, Report 90: New-Product Evaluation Procedures.[44]

Although this section details the procedures for the evaluation of new or unusual noise barrier products, designs, and concepts, it is equally as relevant for recycled material and common noise barrier products from new or existing manufacturers.

If specific requirements are not available or have not been developed by a particular organization, which is particularly typical for recycled materials, then the basic acceptance principles of durability, safety, functionality, life-cycle costs, and cost effectiveness should be thoroughly addressed and proven.

The current trend in writing specifications is leaning toward performance-oriented requirements rather than detailing such things as specific ingredients in a compound or how to manufacture a specific component. This has allowed the manufacturer to be more creative in both the materials used and in the overall design of the product. Generally, this has resulted in a more durable and better designed product with much more competitive prices. This type of specification is also particularly suited for dealing with new design concepts as well as composite, virgin and recycled materials.

- **Material Description** - Every submission should have a detailed description of all materials used in the noise barrier structure including details of the standards to which each component was manufactured. This would encompass everything from the panels, posts, and footings, down to the type of material used for **mandrels** in rivets. If the specifications stipulate restrictions on the materials that can be used, then a statement to this fact should be included in the submission. Otherwise, the manufacturer should be responsible for proving that it would be of benefit to the responsible organization that these restricted materials be used.

If common construction materials are being used, such as concrete, steel, or wood, the specific requirements are probably included in the noise barrier specifications. However, the proliferation of materials such as new plastics and recycled materials has most likely not been addressed in many specifications. Most specifications will have only general references to these, which generates another instance when the acceptance fundamentals, as described before, along with performance criteria, should be used.

- **Test Results** - Test reports are crucial to any acceptance process. How tests are conducted and by whom could mean the difference between acceptance and rejection of the submission. All testing should be done by an accredited, independent laboratory acceptable to the approving responsible organization. However, on some occasions, it may be necessary to have specific testing conducted by the manufacturer. This should only be allowed in the presence of the responsible organization's representative knowledgeable in the specific test procedure.

Whenever possible, testing of samples should be performed using full, production run products. In some cases, it might be advantageous to have the approving organization select the samples for testing. Caution should be taken when the tests are conducted on samples made specifically for testing purposes.

The submission should include complete copies of all required test results. Names of the laboratories, technicians, and telephone numbers should also be available.

- **Drawings and Associated Notes** - The drawings should include the following items:
    - Details of the individual components;
    - Location and sizes of reinforcing bars in concrete (if applicable);
    - Weld details (if applicable);
    - Illustrations of the assembled noise barrier system;
    - Component assembly details;
    - Installation staging and techniques;
    - Location of manufacturer's name and lot number on major components;
    - Footing design details for installation in both earth and rock;
    - Typical grading details;

- Access openings and door details;
- Access openings/valves for fire hoses;
- Structure mounting details;
- Termination and transition treatment;
- Post configuration for changes in alignment; and
- Panel stepping details.

The accompanying notes should include:
- The structural codes to which the noise barrier was designed;
- Reference wind pressure;
- Handling and storage requirements;
- Noise Reduction Coefficient (only for sound absorptive noise barrier products);
- Sound Transmission Class;
- Properties of all materials used such as minimum strengths, type of steel used, and details of coatings; and
- Installation procedures.

All drawings and accompanying notes should be dated, stamped, and signed by a registered engineer.

- **Calculations** - Typically, noise barrier manufacturers are responsible for the structural integrity of their noise barrier system. Therefore, copies of comprehensive design calculations should be provided demonstrating that the panels, posts, and footings are capable of withstanding various wind loads, **ice credations**, and soil conditions in accordance with the codes stipulated by the approving organization.

- **Quality Control Plan** - The quality control plan should outline the manufacturer's procedures to ensure production of a consistently acceptable product. This plan will vary considerably depending on the type of materials used as well as the differences in the manufacturing processes.

- **Maintenance** - Details should be provided to indicate what protection has been used to address the graffiti issue, and, specifically, how to restore the components to their original appearance. Also, in cases of minor damage, a description of the repair technique should be included.

After a review of the information presented in the submission, the first of several decisions should be made with respect to adoption, rejection, or further evaluation of the product.

New product evaluation forms should be used to obtain information from the manufacturer to aid in properly evaluating the new product. These forms assist in providing orderly and concise information, reducing misunderstandings, and avoiding delays resulting from the absence of needed information that may be available but not included in the submittal.

A generic new product evaluation form is typically designed to obtain the following information:
- Trade name of product;
- Manufacturer and location(s);
- Manufacturer's representative and location(s);
- Background of company;
- Date product introduced on market;

- Recommended use and limitations;
- Description, composition, laboratory analysis;
- Plans, sketches, photographs, specifications;
- Patent status;
- Royalty costs;
- Guarantee terms;
- Features and advantages claimed;
- Existing standards, plans, specifications met;
- Instructions for use;
- Material and in-place costs;
- Limitations on availability;
- Willingness to provide samples (cost);
- Willingness to demonstrate;
- Precautions to be used in handling;
- Known health hazards;
- Acceptance status within other agencies;
- Status of trial installations; and
- Names of persons contacted in department.

This form can also be used to notify the manufacturer of any conditions that must be met before evaluation will be considered.

10.1.2 Step 2 - Preliminary Examination.
The second step of the new product evaluation process typically includes an in-depth study of all material submitted by the manufacturer. At this time a needs assessment should be made. Acceptance of products for routine use at the completion of this stage is rare. However, most products offering no more than moderate promise may be rejected at this time.

10.1.3 Step 3 - Detailed Evaluation.
The third step in the new product evaluation process may consist of laboratory testing and/or field trials to test performance under local field conditions. This is likely to be the most time-consuming part of the evaluation process and is applied only to those new products that appear to have a particularly good chance of filling a need. Some caution should be exercised in the acceptance of samples for testing to be certain that the samples truly represent the product to be furnished. Randomly chosen samples are preferable to stock samples. In many cases, the field trials will provide the final evidence that a new proposed noise barrier product will meet established requirements.

Well-designed experiments, careful attention to performance observations, accurate record keeping, and thorough reporting of the results are the essential components of any successful product evaluation system. Improperly designed and conducted experiments that result in early failure can be both expensive and a source of public embarrassment and can also lead to inconclusive results that may unnecessarily prolong testing or, when serving as the basis for rejection of a product, produce controversy between the manufacturer and the responsible organization.

Each material should also be assessed for their ability to be recycled or disposed of in an acceptable and cost effective manner at the end of their useful life. This is particularly critical when considering the use of components already made of recycled materials.

Completion of this stage of the new product evaluation process usually leads to either acceptance or rejection of a product, except in those few instances where the need for further evaluation becomes evident.

10.1.4 Step 4 - Incorporation into Standards, Specifications, Manuals, and Policies.

Once a product has been found to be acceptable, evaluation results should be converted into the "media of practice," which include standards, specifications, manuals, policy statements, and the like.

10.1.5 Step 5 - Implementation.

Manufacturers usually can be expected to apply their promotional expertise in encouraging the use of approved new products. However, internal measures considered to be within the scope of the total new product evaluation process may also be necessary for a new product to reach its full potential of usefulness. Some of these measures may include demonstrations, policy statements, training materials, documentation feedback, workshops, and promotional announcements.

10.1.6 Step 6 - Performance Feedback.

Follow-up observations and evaluations of the in-service performance of approved new products can provide documentation of benefits, point out early failures, and call attention to adjustments that may be needed to achieve intended objectives. To ensure that the evaluation process works, it is essential that a comprehensive regiment of conducting routine observations of service performance be established. It is also important to involve field staff in this stage of evaluation by encouraging them to routinely check for early signs of problems by instructing them on what to look for and how to report it.

Well organized channels of communications within agencies regarding new product evaluation and use are necessary to receive full value from evaluation systems. Well planned evaluations that produce conclusive results and orderly and understandable reports with appropriate documentation have an important bearing on the correctness of decisions regarding the disposal of products that have undergone evaluation.

The regular exchange of information on new products among states and on a national and international level is also valuable for locating useful new products and avoiding unnecessary duplication of evaluation efforts.

It is useful to recognize that the concepts and techniques of new product evaluation are similar in many ways to the concepts and methodologies of value engineering. NCHRP Synthesis 78: Value Engineering in Preconstruction and Construction contains a considerable amount of information on methodology that applies to new product evaluation as well as to value engineering.[45]

Although value engineering concerns a broader area, the basic aims of new product evaluation and value engineering are the same. Both seek to do more with available resources and without loss of service and both apply user-oriented approaches. Further understanding of the similarities can be gained from the following definition of value engineering as provided by the Society of American Value Engineers: "Value engineering is the systematic application of recognized techniques, which identifies the function of a product or service, establishes a value for that function, and provides the necessary function reliably at the least overall cost."

Due to the many similarities between the aims and processes of new product evaluation and the carefully structured and tested methodologies that characterize value engineering, it is recommended that these methodologies be considered in any review of an existing new product evaluation system for the purpose of improvement.

10.2 Product Handling and Storage
Handling and storage requirements should be detailed for all of the various components while at the manufacturing facility, en route to the project, and on the project site.

10.3 Sampling and Testing Requirements
Every product that is installed along highways should be required to meet certain expectations for safety, performance, and durability. Noise barriers are no exception to these requirements. With respect to safety and performance, the requirements and evaluation of noise barriers are specific, and are usually described in great detail by local structural design codes and environmental regulations. However, when it comes to durability, the individual specifying agencies are usually left to develop their own evaluation procedures. Defining an appropriate test method that predicts long-term performance under actual field conditions is the challenging task.

To obtain valid results, the specimens used should be taken from a finished production run product as opposed to small handmade pieces that were specifically made for the test. By insisting on this, the test results can also be used to evaluate in-plant quality control and production operations.

In general, the tests discussed below apply to all types of barrier systems. Additional tests applicable to specific barrier systems (concrete, metal, and wood) follow in Sections 10.3.1 through 10.3.3.

- **Structural Strength** - The noise barrier system should be tested for structural strength properties to determine if it meets local requirement. Strength of the panel should be verified through load testing on a production panel sample. The method of testing may vary depending on the product composition and structure and, to a lesser degree, on regional conditions in which the system will be installed.

- **Shatter Resistance** - Shatter resistance should be considered as both a maintenance issue and a safety issue. If the material is suspected of being fractured or shattered, an analysis of the product's shatter resistance should be conducted to determine if the material's characteristics are acceptable for use in a noise barrier system. A typical test method to determine susceptibility to shatter is ASTM E695.[46] This test method outlines the procedures for measuring the relative resistance of wall, floor, and roof construction when subjected to various types of impact loadings.

- **Flame Spread and Smoke Generated** - To ensure that the retardants are adequate, the general practice is to set the minimum allowable rate of flame spread and smoke generated to a value not greater than the rate for a typical fence material, such as pine. To determine the material's burn characteristics, it should be tested in accordance with ASTM E84, and, through the use of a numeric indexing, the results should be comparable with or better than the results achieved from pine under the same test conditions.[47]

- **Toxicity** - For uncommon or recycled materials, concerns for environmental damage and health hazards should be addressed by requesting appropriate leachate testing or other methods to determine the toxicity of the final noise barrier panel material. The appropriate test methods to be specified are usually governed by the respective environmental organizations.

10.3.1 Concrete.

- **Slump Test (Suitable for Wet-Cast Concrete Only)** - This test determines the stiffness and consistency of freshly mixed concrete and, in general, is a good indicator of the amount of water in the mix. Although water is essential in the production of concrete products, too much water tends to increase the distance between the particles in the mix to a point where they are too far apart to create a good strong link with each other. Not enough water will prevent the mix from hydrating properly and, hence, prevent the chemical reaction needed to develop a strong bond between the particles in the mix.

  Samples should be taken just before pouring into the molds or form work. Slump should be measured in accordance with CAN/CSA A23.2-5C (Methods of Test for Concrete). This test is not applicable for dry cast concrete mix, since it is only slightly dampened and does not tend to deform under it's own weight.[48]

- **Air Content (Suitable for Wet-Cast Concrete Only)** - This test determines the amount of air in cured concrete. The test is a good indicator of durability of concrete which may be frequently exposed to freezing and thawing conditions. The measure of the volume of air entrained and entrapped in the concrete is expressed as a percentage of the total volume of the sample. Samples should be taken just prior to pouring into the molds or form work. Testing should be performed in accordance with CAN/CSA A23.2-4C (Air Content of Plastic Concrete by the Pressure Method).[49]

  This test is not applicable for dry cast concrete mix due to the difference in void structure and consistency of dry cast concrete.

- **Compressive Strength** - This test determines the maximum compressive strength of cured concrete samples. A load is applied uniformly over the entire surface area of the sample ends. The strength is measured at the point of failure and is expressed in MPa (PSI) which represent the average cross-sectional area of the sample.

  There are 2 methods of obtaining samples: One is molding a cylinder from the fresh concrete mixture during the pouring stage of the casting process. The other is by cutting a sample from a cured finished product. The cylinder method is more suitable for wet cast products. The cut sample method is more suitable for dry cast products. Samples should be prepared according to ASTM C684.[50]

No reinforcing bars or mesh should be present in the cut samples since the cutting or coring operation may damage the concrete surrounding the reinforcing when the cutting or coring bit comes in contact with the metal. Testing should be in accordance with ASTM C496.[51]

- **Air Void Analysis (More Suitable for dry-cast concrete products)** - This test determines the shape and size of air voids in cured concrete samples. Testing should be in accordance with ASTM C457.[52]

- **Freeze-Thaw/Salt Scaling** - There are two standard test methods currently available for this test: (1) ASTM C666; and (2) ASTM C672.[53,54]

ASTM C666 determines the resistance of concrete specimens to rapidly repeated cycles of freezing and thawing in the laboratory by two different procedures: A-Rapid Freezing and Thawing in Water, and B-Rapid Freezing in Air and Thawing in Water. This method is not an appropriate test for noise barriers on the basis that, by its own admission, the standard states that "Neither procedure (A or B) is intended to provide a quantitative measure of the length of service that may be expected from (any) type of concrete." In addition, only distilled water is used during the procedure. This does not represent common field conditions in northern or coastal regions where these products are constantly subjected to salt laden moisture.

A number of other concerns related to ASTM C666 test procedures have been documented by the Strategic Highway Research Program, in a 1994 report (SHRP-C-391) identifying problems with the design of the apparatus used for the test as well as the inadvertent drying of the samples during the air freezing cycle.[55]

ASTM C672 can be used on all concrete products, both precast and cast-in-place, to evaluate the effects of air content, cement factor, slump, water-to-cement ratio, surface treatment, curing and other variables on concrete's resistance to salt scaling and rapid freezing and thawing. It should be noted that some agencies have adopted a slightly modified version of this test method. The modifications include:
- The use of a 3 percent sodium chloride solution for both the preliminary conditioning of the sample and the actual test itself; and
- A quantitative method of measuring deterioration during the test period.

Example:
After 50 freeze/thaw cycles, the loss of mass from the surface of any sample shall not exceed 0.8 kg per m^2, and the samples shall exhibit no deterioration in the form of cracks, spalls, delamination, aggregate disintegration, or other objectionable feature.

- **Density** - Determining the density of the concrete material provides information related to the degree of compaction the concrete mix was subjected to in the mold prior to curing. The denser the product, the better the quality of concrete assuming that a suitable mix design was used and the product was cured properly.

- **Water Absorption** - This test determines the amount of water the sample can absorb over a given time period. Generally, the more water absorbed, the poorer the quality. Samples should be cut from the cured concrete product. By thoroughly drying the samples in an oven and then submerging into a tank of distilled water for a specific length of time, the percentage of absorption can be determined. Since this is a relatively common test with each responsible organization having their own preferences, this section will not suggest any one method to be used.

- **Minimum Cover Over Reinforcing** - Panels and posts should be checked to ensure that the minimum concrete cover over the reinforcing is maintained during the casting operation. Adequate cover is critical in preventing premature penetration of salt laden moisture from reaching the reinforcing bars. This results in the corrosion of the bars and subsequent spalling of the concrete surface along with the drastic deterioration of the structural properties of the components.

  There are several methods available to determine the amount of concrete cover over reinforcing bars of mesh. The most accurate method (but also the most destructive) is by cutting through the concrete to expose its core and to physically measure the distance from the concrete surface to the reinforcing bars. A less destructive method is to use specialized x-ray and electronic equipment to detect the location and, depending on the technology used, the amount of cover and the size of the bar or mesh. Although these devices are, to varying degrees, less accurate than the cutting method, it is non-destructive and can usually be performed in-situ.

- **Dimensions** - All precast and cast-in-place concrete products should be checked for proper dimensioning (see Section 11.5.1).

- **Visual Inspection** - All precast and cast-in-place concrete products should be visually examined to identify any unusual and unwanted features which will affect the structure, durability, and performance of the noise barrier wall, such as honey combing, knuckling, cracks, and voids.

10.3.2 Metals.

- **Accelerated Weathering** - This test provides information on how well the metal and its coating withstand extreme weather conditions. Although this accelerated test method is not a true representation of actual condition, it does provide a reasonable tool to predict the results of long-term exposure to harsh climatic conditions. ASTM B117 is a suitable test method and has been thoroughly quantified over the numerous years in practice.[56]

- **Coating Durability** - The coating system should be tested in a weatherometer chamber in accordance with ASTM G26.[57] The coating system should then be evaluated for the following weathering effects when rated according to the appropriate ASTM standards:
  - Checking - ASTM D660;[58]
  - Cracking - ASTM D661;[59]
  - Blistering - ASTM D714;[60]
  - Adhesion - ASTM D3359;[61]
  - Color change - ASTM D2244;[62] and
  - Chalking - ASTM D4214.[63]

- **Coating Thickness** - Coating thickness, whether it is galvanized, painted, sprayed, or dipped, must be verified to ensure compliance with specifications.

10.3.3 Wood.

- **Structural Grade** - Specifying a good structural grade of lumber does not guarantee that all pieces of wood will be straight enough to permit the tight fit normally required for wood barriers. Therefore it is essential to visually confirm the grade of the wood used. In addition, any boards that are warped, checked, split, or have excessive knots should be removed.

- **Dimensional Stability** - Structural graded lumber does not ensure that the product will never shrink, particularly if the wood has not been properly seasoned or kiln dried before pressure treating. Therefore, all wood components, particularly timbers beams, and posts thicker than 100 mm (4 in), should be checked for dimensional tolerances (see Section 11.5.1).

- **Determination of Penetration** - AWPA A19 is a suitable test to determine the depth of penetration of the preservative into the wood.[64] The penetration rate may vary between species.

- **Moisture Content** - ISO 4470 details a method used to determine the amount of moisture in the wood.[65] It is a nondestructive test and should be conducted on all lots, individual larger pieces, and particularly those where warping, checking, and splitting may not be preventable or may have a serious impact on the overall performance of the wall.

10.4 Criteria for Approval

This section will discuss the more common procedures for submissions and evaluation leading up to the acceptance of a noise barrier installation. Further guidance can also be obtained from CAN/CSA Z107.9.[66] Final acceptance of a noise barrier installation can typically be broken down into four major segments:

- Acceptance of the noise barrier system design (see Section 10.4.1);
- Acceptance of the noise barrier manufacturer (see Section 10.4.2);
- Acceptance of the site specific shop drawings (see Section 10.4.3); and
- Acceptance of the installation (see Section 10.4.4).

10.4.1 Acceptance of the Noise Barrier System Design.

Common practice includes submission, by the owner of the noise barrier system design, for evaluation by the responsible organization. The submission normally would include detailed generic or non-site specific information of a particular noise barrier system. See Section 10.1.1 for details.

Acceptance of a proprietary noise barrier system should be based not only on the test data and documentation submitted with the application but also on subsequent results of laboratory and field testing and trial installations. Further, acceptance should be granted only if the design and materials conform to the required specification and that the system has been demonstrated to be constructable.

General grounds for a noise barrier system being accepted or rejected could be based on, but not limited to, any of the following conditions:

- Unsatisfactory field performance;
- Has the potential for or creates environmental damage;
- Has the potential for or creates a health hazard;
- Failure of the material to comply with any parts of the prescribed specifications; and
- Inaccurate information or claims contained in the submission.

If the vendor changes the design of the system or any of its components, then the system should be reevaluated.

10.4.2 Acceptance of the Noise Barrier Manufacturer/Fabricator.

The manufacturer/fabricator of any or all noise barrier systems components should be required to provide information related to the production of the noise barrier system. If, after acceptance, the manufacturer initiates any variations to the system design or any changes to component fabrication and/or plant location, reevaluation and possible re-qualification of the manufacturer/fabricator should be considered.

All submissions should at least include the following information:
(1) The trade name of the product;
(2) Manufacturer/fabricator name and address;
(3) Plant location(s);
(4) List of supplier(s);
(5) Quality control/quality assurance program(s) including:
   a. Methods of storing raw and finished materials;
   b. Mixing, batching, and/or assembly controls;
   c. Inspection, testing methods, and frequency;
   d. Details of production lot test certification;
   e. Product identification marking details;
   f. Packaging, shipping, and handling requirements; and
   g. References and manufacturing history along with contact names and addresses.
(6) All test reports and certifications be dated no more than one year prior to submission;
(7) Drawings and calculations be signed, sealed, and dated by a professional engineer licensed in the jurisdiction for which the approval is being sought;
(8) When possible, the manufacturer or supplier should provide advance notice to the owner of the data that the fabrication of the noise barrier material is expected to commence; and
(9) All materials delivered to the construction site be visually inspected by the owner for proper dimensions, honeycombing, cracks, voids, surface defects, inconsistency in color and texture, and any other damage or imperfections.

Acceptance of the noise barrier manufacturer/fabricator should be based not only on the test data and documentation submitted with the application but also on subsequent results of laboratory and field tests trial installations. Further, acceptance should be granted only if the design and materials conform to the required specification, and the manufacturer has demonstrated the ability and has the equipment and facilities necessary to consistently produce an acceptable product.

Free access should be allowed to the place(s) of manufacture of the noise barrier components for the purpose of random inspection and examination of plant quality control records, certificates, materials used, process of manufacturing including, but not limited to, welding, galvanizing, prefabrication, and precasting, and to make any test as may be considered necessary to ensure compliance with this specification.

General grounds for a noise barrier manufacturer/fabricator being accepted or rejected could be based on, but not limited to, any of the following conditions:

- Unsatisfactory field performance;
- Has the potential for or creates environmental damage;
- Has the potential for or creates a health hazard;
- Failure of the material to comply with any parts of the prescribed specifications; and
- Inaccurate information or claims contained in the submission.

10.4.3 Acceptance of Project Specific Design Details (Shop Drawings and Related Documents).

The vendor of the noise barrier system should submit project or site specific design details (shop drawings and documents) for each system to be installed. If, after acceptance, there are any changes made to the system design or to component fabrication and/or plant location, reevaluation and possible re-qualification of the noise barrier system and the manufacturer/fabricator should be considered.

The submission shall include at least the following information:
(1) The trade name of the product;
(2) Name and address of the noise barrier system owner (if different than the manufacturer/fabricator);
(3) Manufacturer's/fabricator's name and address;
(4) Plant location(s);
(5) List of material supplier(s);
(6) Full engineering, and fabrication drawings;
(7) Site specific installation details including but not limited to:
   a. End treatment;
   b. Elevations;
   c. Stepping of sections;
   d. Connections to existing structures;
   e. Consideration for above and underground utilities;
   f. Footing design;
   g. Structural calculations; and
   h. Aesthetic and landscaping treatment details.
(8) All reports and certifications be dated no more than one year prior to submission; and
(9) Drawings and calculations be signed, sealed, and dated by a professional engineer licensed in the jurisdiction for which the acceptance is being sought.

Acceptance of the noise barrier project or site-specific design should be based on, but not limited to, the reports, drawings, and documentation submitted with the application as well as subsequent results of laboratory and field tests. Further, acceptance should be granted only if the design and materials conform to the required specifications.

A noise barrier system project could be rejected if it demonstrates any of the following conditions:

- Unsatisfactory field performance;
- Has the potential for or creates environmental damage;
- Has the potential for or creates a health hazard;
- Failure of the design, material, or components detailed in the submission to comply with any parts of the prescribed specifications; and
- Inaccurate information or claims contained in the submission.

10.4.4 Acceptance of the Installation. The installer of the noise barrier system should provide a noise barrier system using acceptable manufactured/fabricated components in accordance with project specifications (shop details). The installer should provide to the responsible organization, prior to installation, all test and production certificates for each production lot supplied by the manufacturer/fabricator, showing compliance with all requirements of the prescribed specifications. Any variations or changes during installation from the accepted noise barrier system design, the accepted component fabrication and/or plant location, or the accepted project specific details should warrant reevaluation of the installation and possible re-qualification of the noise barrier system, the manufacturer/fabricator, and/or the project specific details.

Acceptance of the noise barrier installation should be based on compliance with all the requirements and specifications associated with the noise barrier system as detailed by the design, the manufacturer/fabricator the project specific details, and any other prescribed specifications, as well as plant inspections and subsequent results of laboratory and field tests prior to, during, or after installation.

If any part of the installation fails to comply with any of the prescribed requirements, the noise barrier installation should be corrected by the installer so that all components meet those specifications. Further, any components damaged during installation or alterations should be disposed of and replaced by the installer.

A noise barrier installation should be rejected if it demonstrates any of the following conditions:

- Unsatisfactory field performance;
- Has the potential for or creates environmental damage;
- Has the potential for or creates a health hazard; and
- Repeated failure of the design or the material to comply with any parts of the prescribed specifications.

Section Summary

Item #	Main Topic	Sub-Topic	Consideration	See Also Section	✓
10-1	Evaluation Process	Submission	It is critical to know what the requirements and specifications are for acceptance by the responsible organization. And, just as important, in what format are the agencies accustomed to seeing these submissions?	10.1.1	
		Detailed Evaluation	Some caution must be exercised in the acceptance of samples for testing to be certain that the samples truly represent the product to be furnished.	10.1.3	
		Performance Feedback	It is essential that a comprehensive regiment of conducting routine observations of service performance be established. It is also important to involve field staff by encouraging them to routinely check for early signs of problems, by instructing them on what to look for, and how to report it.	10.1.6	
10-2	Sampling and Testing Requirements		The testing specimens used should be cut from a finished production run product as opposed to small handmade pieces that were specifically made for the test. By insisting on this, the test results can also be used to evaluate in-plant quality control and production operations.	10.3	
10-3	Criteria for Approval	Acceptance of Design	If the vendor changes the design of the system or any of its components, then the system should be reevaluated.	10.4.1	

11. INSTALLATION CONSIDERATIONS

The noise barrier system, including the posts, panels, and foundation, must be designed and installed to withstand local wind loads as determined by the reference wind specifications for each unique installation, in conjunction with the soil design parameters along the centerline of the wall. Regardless whether the noise barrier system is to be prefabricated or fabricated on site, the manufacturer/ contractor should be required to provide advance notice of the date that the fabrication of the noise barrier components is expected to commence. This is to provide sufficient opportunity for the responsible organization to observe and, possibly, take samples for quality control and quality assurance testing.

This section discusses a variety of installation-related processes including:

- Site grading and preparation (Section 11.1);
- Foundation requirements (Section 11.2);
- Noise barrier manufacturing;
- Quality assurance (Section 11.3);
- Handling and storage of materials (Section 11.4);
- Barrier assembly and construction (Section 11.5); and
- Construction noise barriers (Section 11.6).

11.1 Site Grading and Preparation

Earth grading and/or paving at the base of the noise barrier panels and posts should be such that the bottom panels are effectively buried to avoid any gaps. The earth and/or pavement should be sloped away from the installation to prevent the washout of soil at the base of the noise barrier.

All graded earth should be sufficiently compacted in accordance with ASTM D1557.[67] This is to reduce or eliminate settling of the surrounding earth and possibly creating gaps at the base of the wall between the panels and the ground line.

Changes in vertical alignment, or the ground line, should occur at the posts, or the posts should be installed where these changes occur.

11.2 Foundation Requirements

The site-specific type, depth, size, and shape of concrete footing foundations should be determined in accordance with the governing codes based on the determined soil design parameters along the alignment of the noise barrier installation. The topic of footings was discussed previously in Sections 8.4.1 and 8.4.2

11.3 Quality Assurances

All materials delivered to the construction site should be checked to ensure that they meet all the stipulated requirements. This is typically done in three ways:

- Visual examination (Section 11.3.1);
- Proof of certification (Section 11.3.2); and
- Testing of samples either on or off site (Section 11.3.3).

11.3.1 Visual Examination.
A visual examination of all components should be done at the plant and at the site before, during, and after installation. Specifically, components should be visually inspected for:

- Proper dimensions;
- Surface defects;
- Inconsistency in color and texture;
- **Distress**;
- Cracks;
- **Coating holidays**;
- Rust stains;
- Deterioration;
- Warping;
- Honeycombing (for concrete barrier systems);
- Voids (for concrete barrier systems);
- Spalling (for concrete barrier systems);
- Delamination (for concrete barrier systems);
- Oxidation (for metal barrier systems);
- Checking (for wood barrier systems);
- Splitting (for wood barrier systems);
- Loose or missing knots (for wood barrier systems);
- Lack of **incising** (for wood barrier systems); and
- Any other surface damage or imperfections such as dents, scratches, chips, and gouges.

11.3.2 Proof of Certification.
In some cases, the quality of a product may be inspected by another approving or certifying responsible organization, or a self-governing industrial association. A certificate may be provided from the manufacturer of that particular component, certifying that the product had been manufactured to a specific standard and that the materials used are of a certain quality. This type of quality assurance certification is typical for most steel products where the certificate is issued by the mill, usually certifying the grade of the metal and some physical and chemical properties. These certificates should be made available at the time of supply.

11.3.3 Testing Methods.
There are two types of physical testing that can be done on any product: (1) destructive; and (2) non-destructive. **Nondestructive testing** allows the product to be evaluated without destroying its properties or appearance. If this type of testing cannot provide the necessary degree of accuracy or adequate information, some form of destructive testing may be required. There are varying degrees of destructive testing, ranging from total destruction of the product to minor cosmetic damage, which may be either too small to see or easily repairable.

11.3.4 Sampling.
Samples for additional testing should only be provided when products are delivered to the site, which are ready for installation and have passed both the visual inspection and the proof of certification requirements.

11.4 Handling and Storage of Materials on Site
Handling and storage of noise barrier components on site should be according to the manufacturer's specifications.

11.5 Barrier Assembly and Construction
This section discusses the considerations associated with barrier assembly and construction, including construction tolerances, sealants, fasteners, and other related processes.

11.5.1 Construction Tolerances.
To ensure proper fitting of components and an acceptable overall appearance, the noise barrier should be constructed to the specified height and horizontal and vertical alignment within specified tolerances. Regardless of the aesthetic treatment and changes in elevation to accommodate vertical and horizontal alignment shifts, the minimum specified height of the noise barrier should be maintained at all times.

11.5.2 Grout, Caulking, and Sealants.
Any gaps or cracks in the installed noise barrier system should be filled with an acceptable filler materials (i.e. grout, caulking, backer rods, or sealant). When this type of material is recommended, it is important that all surfaces be clean, dry, and free of loose material before applying the filler material. In some cases, it may be necessary to use a flashing to cover larger gaps or to redirect water away from the noise barrier wall. In all cases, it is critical to ensure that light cannot be seen through any part of the noise barrier system and that the filler material or **flashing** does not trap water inside the wall components or near the base and foundation of the wall.

11.5.3 Anchors, Fasteners, and Lifting Inserts.
Anchors, fasteners, and lifting inserts should be either made of a non-corroding metal or have a corrosion-resistant coating applied after fabrication, but before installation.

11.5.4 Field Welds.
Factory-applied coatings, such as paints, stains, or galvanizing, are considered to be far superior to field-applied methods. Therefore, any field operation that would require recoating the component surfaces, such as welding of metal components, should be kept to an absolute minimum to maintain the integrity and durability of the original coating. If field welds are necessary, the weld and its surrounding area should be cleaned and painted with an acceptable organic zinc-rich paint matching the color of the surrounding surfaces.

11.5.5 Coatings.
If painted, stained, or galvanized surfaces become abraded during shipping, storage, or installation, those surfaces should be cleaned and re-coated with an organic zinc-rich paint that matches the color of the surrounding surfaces.

11.5.6 Installation Jigs. During the installation of noise barriers, devices are often used to support barrier system elements (see Figure 236). While form work is typically used to support cast-in-place concrete pours, installation jigs are devices which are assembled and placed to hold pre-assembled or pre-manufactured components such as noise barrier posts and/or panels. A typical use of a jig may be to hold a noise barrier post at its final position in an excavated foundation while the foundation concrete is poured and during the concrete's curing process. Jigs may also be used to support noise barrier panels during the process of anchoring the panels to their supporting structure.

Figure 236. Installation jigs (data base #6536).

11.5.7 Installation Scaffolding. In most cases, scaffolding is needed to install most types of noise barriers. Cranes may be used to install prefabricated or preassembled panels (see Figure 237), but crews are still needed on scaffolding to fasten the panels to the posts and framework (see Figures 238 and 239). Scaffolding needs room, a good solid foundation, and a considerable amount of effort and time to install. All of these factors should be considered before this type of material is selected for a specific site.

Figure 237. Installation considerations (data base #6535).

Figure 238. Installation considerations (data base #2406).

Figure 239. Installation considerations (data base #2327).

11.6 Construction Noise Barriers

In certain instances, noise generated by construction activities may warrant the consideration of noise abatement during the project construction phase. Such abatement may be necessitated by the proximity of construction activities to sensitive receptor sites, the activities at the receptor site itself, the time of both construction and receptor site activities, and the type of construction activity. It may be possible to limit working hours, limit certain activities during certain hours, and/or adjust project construction staging to allow for the earliest possible construction of the project's permanent noise barriers. Figure 240 shows a permanent noise barrier built prior to much of the project's construction activities. It may also be necessary to provide temporary noise barriers for certain operations. On any particular project, it may be necessary to use varying combinations of any or all of these techniques.

Figure 240. Construction noise barriers (data base #6539b).

11.6.1 Temporary Noise Walls and Berms.

Both stationary and mobile noise sources may be shielded by placing either a noise berm or wall between the source and the receiver. Such a berm, if needed for only a relatively short period of time, may be constructed of virtually any construction material, with minimum stabilization (as per soil erosion requirements) provided. Noise wall enclosures can be constructed of material such as construction grade plywood or sheet metal. They may be built in sections which can be easily transported to the location needed, assembled, disassembled after use, and moved for use at a new location. It may be possible to create a construction noise barrier by temporarily supporting noise panels which will ultimately become part of a permanent noise barrier system.

11.6.2 Early Construction of Permanent Noise Walls and Berms.

If possible, the desirable construction noise abatement technique for projects having permanent noise barriers is to construct such barriers as early in the construction process as possible. The barrier can then be used to abate construction activity-related noise and, ultimately, the highway traffic noise it is designed for. With this scenario, more opportunities also exist for placement of stationary noise sources such as compressors in shielded locations. Obviously, such barrier construction may not always be possible as early in the project staging as may be desired. Certain grading, earthwork, fabrication, and drainage operations often must precede such barrier construction. Figure 241 shows an early-constructed permanent barrier protecting adjacent residences from drilling activities.

Figure 241. Early construction of permanent noise barriers (data base #6538).

Section Summary

Item #	Main Topic	Sub-Topic	Consideration	See Also Section	✓
11-1	Site Grading & Preparation		All graded earth should be compacted to reduce or eliminate settling of the surrounding earth and possibly creating gaps at the base of the wall between the panels and the ground line.	11.1	
11-2	Foundation Requirements		The site-specific type, depth, size, and shape of concrete-footing foundations should be determined based on the determined soil design parameters.	11.2	
11-3	Quality Assurances	Testing Methods	Non-destructive testing is the preferred method; however, it may not always be the most accurate, or possible, for the determination and verification of some properties.	11.3	
11-4	Barrier Assembly and Construction	Construction Tolerances	Regardless of the aesthetic treatment and changes in elevation to accommodate alignment shifts, the minimum specified height of the noise barrier should be maintained at all times.	11.5.1	
		Grout, Caulking, and Sealants	It is critical to ensure that light cannot be seen through any part of the noise barrier system and that the filler material or flashing does not trap water inside the wall components or near the base and foundation of the wall.	11.5.2	
		Coatings	If painted, stained or galvanized surfaces become abraded during shipping, storage, etc., those surfaces should be cleaned and re-coated.	11.5.5	
11-5	Construction Noise Barriers		It may be necessary to provide temporary noise barriers or begin early building of permanent noise barrier depending on the magnitude of the construction activities for a particular project.	11.6	

12. MAINTENANCE CONSIDERATIONS

The expected maintenance-free life span of a noise barrier varies considerably based on many factors, including the barrier material type, surface texture, color, and component parts. Climatic conditions and the barrier's relationship to the roadway also play a role in the durability of most barriers. Many of these issues have been discussed elsewhere in this manual. Additional considerations are discussed below.

12.1 Repairs

Noise barriers will become damaged at some point in their life, either from handling mishaps during construction, installation defects that appear well after the barrier has been installed, vehicles or debris hitting the wall, or simply from old age and exposure to the elements over time (see Figures 242 to 245). The reasons to initiate a repair may vary depending on the severity and extent of the damage and the policies of the governing responsible organization. In all cases, the following factors should be considered in the decision to repair. The relative weight that each of these factors is assigned may vary locally. However, the order of importance should not deviate significantly from the order in which they are presented.

(1) **Safety** - Is the damage severe enough that the structural integrity of the barrier has been compromised; or have components been repositioned to create an obstacle/hazard to vehicles or pedestrians?

(2) **Durability** - Is the damage severe enough to diminish the durability or life expectancy of some or all the noise barrier components?

(3) **Performance** - Is the damage severe enough to significantly reduce the attenuation provided by the noise barrier system; or, in the case where the barrier also acts as a fence, is it possible for someone to have access through the wall?

(4) **Aesthetics** - Is the damage severe enough to create an unsightly appearance that is deemed unacceptable by the neighboring community?

Figure 242. Repairs (data base #736a).

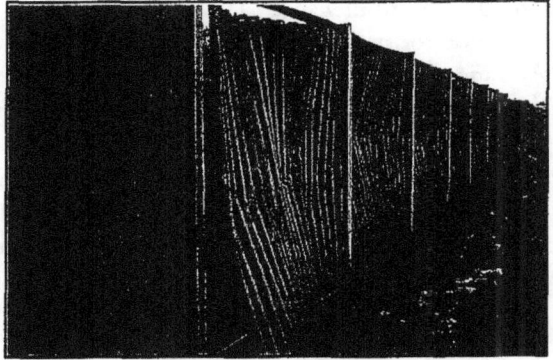

Figure 243. Repairs (data base #2488).

Figure 244. Repairs
(data base #1052).

Figure 245. Repairs
(data base #450).

12.2 Availability of Replacement Parts

Since replacement of barrier elements may be required throughout the life of the noise barrier, the availability of replacement parts becomes a critical issue. If the components are standard products, such as steel "I" beams for posts, this issue will have very little importance. However, if the components are custom made for a specific project, then the issue becomes very critical (see Figure 246).

To address this concern, some agencies have instituted a stock piling policy where the contractor/manufacturer, at the time of construction, supplies additional components to the organization

Figure 246. Availability of replacement parts (data base #226).

responsible for maintenance. Typically, an additional 10 percent would by supplied to the responsible organization for stock piling purposes. The disadvantage to this type of practice is that the responsible organization, after several years of constructing noise barriers, may end up with an excessively large amount of varying stock on hand that has the potential of never being used.

The issue of future availability becomes even more critical when the components have to be custom fitted with either very few or none of the pieces the same. In this situation, stock piling may not be an option. This consequence should be seriously considered during the design stage and should be avoided if at all possible.

12.3 Access

Section 9.4 discussed a variety of means for providing access to both sides of noise barriers for general maintenance and emergency access purposes. These requirements exist for both ground-mounted and structure-mounted noise barriers. The placing of a noise barrier is usually dictated by the results of acoustical analyses which are aimed at determining the best location to block line-of-sight between the noise source and the receivers. This location may or may not be the most accessible from either a construction or maintenance standpoint. These accessibility issues should be considered in the design phase in conjunction with the acoustical, construction, maintenance, and barrier material selection issues. If the only location to place an effective noise barrier is relatively inaccessible, then the design should focus on developing a barrier and related surface treatment and landscaping which is relatively maintenance free.

12.4 Surface/Material Wear and Deterioration

All noise barrier materials will wear over time. The severity is dependent on the type of material, proximity to the roadway, exposure to deicing chemicals, climate, and component design. Typical damage associated with wear are:

- **Deterioration from Moisture** - Rusting, usually caused by excessive exposure to moisture and road salt, will slowly degrade or penetrate metal coatings such as galvanizing and paints (see Figure 247). The typical repair is to remove the rust and repaint with a zinc rich paint. Also, if the design of the components promotes the ponding of moisture on its surface, consideration should be given to redesign replacement parts to eliminate this. If left unattended, the metal components of a barrier will continue to rust until there is no metal left. At that stage, the only repair option is to replace the affected components. This type of aging initially affects the aesthetics of the barrier but, if left unchecked, could compromise safety and performance.

  In some cases and climatic conditions, surfaces of barriers may never completely dry or may require extended periods of dry weather before drying. The resulting mildew and mold can compromise the aesthetic appearance of the barrier (see Figure 248). In such areas, consideration should be given in the design phase to selecting a barrier and surface treatment which can minimize such conditions and be capable of being cleaned on a regular basis.

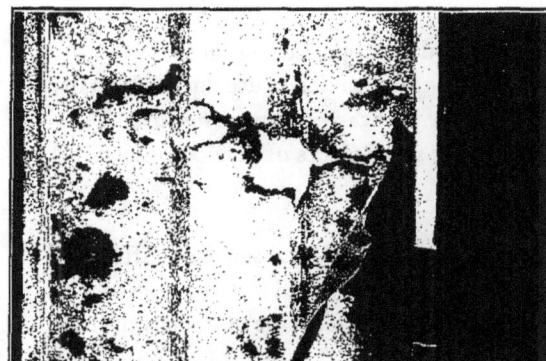

Figure 247. Deterioration from moisture (data base #449).

Figure 248. Deterioration from moisture (data base #1008).

- **Deterioration from Ultraviolet Light Exposure** - The effects of ultraviolet light are most often related to deterioration of the noise barrier's pigments and colors as opposed to its structural elements (see Figure 249). Paints, stains, graffiti coatings (both pigmented and clear), stenciled designs, and integral colors are subjected to fading over time. When using such aesthetic and/or protective treatments, consideration should be given to the eventual maintenance tasks required to retain the desired appearance and/or protection. Consideration of the effects of ultraviolet light is especially critical in the design of transparent barriers as discussed in Section 5.5.1.

Figure 249. Deterioration from ultraviolet light (data base #1501).

- **Loss of Stains and Painted Coatings** - Loss of stains and painted coating can usually be attributed to the sand blasting or small chipping effect caused by roadway dirt and debris being picked up by the air turbulence from fast-moving vehicles. This is more pronounced the closer the barrier is to the roadway.

The damage is usually only to the stain or coating and is typically an aesthetic issue. However, if the coating or stain is providing protection for the barrier material, then it becomes an issue of reduced life expectancy of the material.

The typical repair is to remove any loose material from the surface and repaint or stain with an appropriate durable finish to match the surrounding colors.

- **Warping** - Although most common with wooden noise barrier products, warping (commonly called "**oil canning**") can also occur with metal and concrete panels.

With wood, the common cause is excessive moisture in the wood and improper curing causing the warping. In some cases, it can also be attributed to improper fastening. This defect usually creates gaps or cracks in the noise wall affecting its performance to varying degrees. If the openings are small enough, the repairs could probably be done with caulking material or small pieces of plywood sheeting. Otherwise, the affected components may have to be replaced.

With respect to the oil canning of metal sheeting, warping is usually caused by the absence of stiffening features in the panel profile and is only evident after the panels have been subjected to stresses associated with normal temperature changes. This is a design flaw that normally only affects the aesthetics of the barrier and rarely jeopardizes its safety, durability, and performance.

Concrete warping is more noticeable in stacked panel systems. While full height panels also warp, such warping is not normally visible.

12.5 Landscaping

Comprehensive discussion of landscaping, its relationship to other barrier elements, and its relationship to maintenance factors is contained in Section 6.2. The issue of consistency between the barrier's aesthetic treatment, including landscaping, and the maintenance philosophy of the owner of the barrier, as discussed in Section 6.2, is critical and bears repeating here. No matter how well designed and coordinated a landscape plan may be from the aesthetic standpoint, it is only as good as the ability of the responsible organization to adequately maintain it. It is a waste of time and money to design an aesthetic treatment for which there is neither the commitment (in terms of manpower) nor the funding (long term) to adequately maintain. No matter what the desire from an aesthetic standpoint, the landscape plan needs to be responsive to these financial and manpower constraints. Such constraints may appropriately lead to the selection of vegetation that is native "maintenance free" and to a plan that will foster growth of natural vegetation.

12.6 Graffiti

Section 5.9.3 discusses the various maintenance aspects of the varieties of coatings, stains, and anti-graffiti coatings available for application on noise barriers (see Figure 250).

Figure 250. Graffiti (data base #757).

12.7 Litter

Barrier design should consider the location of a noise barrier in terms of litter susceptibility (Is it a high litter area?), the barrier's ability to "trap" litter, and the philosophy of the responsible organization regarding cleanup of litter (see Figure 251). Often, leaves, grass clippings, and other litter tend to be dumped in the area between the barrier and the right-of-way fence ("no man's land" or "dead man's zone"). If possible, allow adjacent residents to extend their sideline fences to the barrier. Also, landscaping in a high litter area should consider what type of vegetation is best to use. A thorny type of bush may make litter cleanup more difficult than such litter removal from a grassy area. Special design features such as insert areas, planter boxes, etc., may "catch" litter or even become target areas for litterers.

Figure 251. Litter (data base #1642).

12.8 Snow Storage

When noise barriers are installed too close to the roadway it becomes difficult to store snow on the side of the road (see Figure 252). If the snow is piled up on the shoulders, it is a potential safety issue in that vehicles cannot safely pull-off of the main travel lanes during emergencies. This condition usually warrants timely removal of the snow, typically with a front end loader and dump trucks. The removal operation creates serious safety problems with heavy equipment operating so close to the roadway.

The close proximity of the noise barrier will also make it susceptible to damage from snow plowing operations due to the force of the snow being thrown against the barrier and the resultant pressure of the snow piled-up against the barrier.

Figure 252. Snow storage (data base #5342).

12.9 Snow Drifting

Depending on the height, proximity to the roadway and orientation related to prevailing winds, snow drifting may occur across the road as a result of the construction of a barrier (see Figure 253). Drifting creates not only safety problems but also difficulties for snow removal operations compounding the problems associated with snow removal.

Consideration should be given to this concern at the design stage so that, in critical areas, wider shoulders, relocation of the barrier, or possible reduction in barrier heights can be incorporated.

Figure 253. Snow drifting (data base #2540).

12.10 Issues Related to Specific Barrier Types

Special or unique barrier types may sometimes have unique maintenance-related issues which should be considered in the design process. Barriers with large caps (see Section 6.1.3) or special barrier tops (see Section 3.5.5.3) may shade the top portion of a barrier and prevent the natural cleansing of that area by rain water (see Figure 254). Barriers in a "zig-zag" configuration (see Sections 3.5.5.2 and 4.1.2.3.1) present opportunities for plantings within the barriers

Figure 254. Issues related to specific barrier types (data base #3125).

"pockets" or recesses, but may make mowing operations more difficult. Planted noise barriers (see Section 4.1.2.3.2) and noise barriers constructed behind the top of a retaining wall (see Section 4.2.2) may require irrigation and protective fencing (to prevent unauthorized access and climbing). Barriers mounted on structures (see Section 4.2.1) may create special access conflicts if utilities such as electric, gas, fiber optic lines, water lines, or sewer lines are suspended from the bridge or contained in conduits within the bridge beams or parapets.

Section Summary

Item #	Main Topic	Sub-Topic	Consideration	See Also Section	✔
12-1	Availability of Replacement Parts		If barrier components are custom made for a specific project, then the issue of replacement parts becomes very critical.	12.2	
12-2	Access		If the only location to place an effective noise barrier is relatively inaccessible, then the design should focus on developing a barrier and related surface treatment and landscaping which is relatively maintenance free.	12.3	
12-3	Surface/ Material Wear and Deterioration	Deterioration from Moisture	If the design may result in moisture ponding on its surface, consideration should be given to redesign replacement parts. In some climatic conditions, consideration should be given in the design phase to selecting a barrier and surface treatment which can minimize mildew and mold growth and be capable of being cleaned on a regular basis.	12.4	
		Deterioration from Ultraviolet Light Exposure	When using paints, stains, graffiti coatings, and stenciled designs, consideration should be given to the effects of ultraviolet light, especially in the design of transparent barriers.	12.4	
		Loss of Stains and Painted Coatings	If the coating or stain is providing protection for the barrier material, then it becomes an issue of reduced life expectancy of the material.	12.4	
12-4	Landscaping		The issue of consistency between the barrier's aesthetic treatment, including landscaping, and the maintenance philosophy of the owner of the barrier is critical. Manpower and financial constraints may appropriately lead to the selection of vegetation that is native "maintenance free" and to a plan that will foster growth of natural vegetation.	12.5	
12-5	Litter		Barrier design should consider the location of a noise barrier in terms of litter susceptibility, the barrier's ability to "trap" litter, and the philosophy of the responsible organization regarding cleanup of litter. Landscaping in a high litter area should also consider what type of vegetation is best to use. A thorny type of bush may make litter cleanup more difficult than such litter removal from a grassy area. Special design features such as insert areas, planter boxes, etc. may "catch" litter or even become target areas for litterers.	12.7	
12-6	Snow	Storage	Consideration must given to a barrier's susceptibility to damage from snow ploughing operations by both the force of the snow being thrown against the barrier and the resultant pressure of the snow piled up against the barrier.	12.8	
		Drifting	Consideration should be given to snow drifting in the design stage so that, in critical areas, the possibilities of wide shoulders and minimum necessary barrier heights can be incorporated.	12.9	

13. COST CONSIDERATIONS

Because of the vast number of diverse factors which affect the cost of manufacturing, transporting, and erecting noise barriers and their components, no attempt is made in this Handbook to provide specific costs for a particular noise barrier type. However, numerous sources exist which can assist a designer and/or decision-maker in evaluating the cost implications of different barrier types. Two primary sources suggested for initial contact are:

- State and Provincial contacts included in Section 16.5 of this Handbook. These individuals have access to information or sources related to noise barriers constructed in their jurisdiction.
- The *Summary of Noise Barriers Constructed* document[37] published on a regular basis by the U.S. Department of Transportation, Federal Highway Administration, Office of Natural Environment. This document gives details related to barrier type, dimensions, location, cost, etc. constructed in each State.

These initial contacts are likely to provide additional cost-related information sources and contacts such as:

- Specific reports prepared by States or Provinces related to barrier costs;
- Specific project-related information;
- Contractor contacts;
- Supplier contacts; and
- Manufacturer contacts.

The remainder of this section describes additional cost considerations.

13.1 Relationship of Barrier to Project Type

13.1.1 Noise Barrier Built as a Component of Large Construction Project.
The cost of noise barrier construction is usually less if the barrier is built as part of a large construction project. Costs for items, such as mobilization, insurance, maintenance, protection of traffic, etc., are spread out amongst all project elements. The equipment and manpower required for noise barrier erection may already be required for construction of other project elements. In general, such economy of scale factors are present on these large projects.

13.1.2 Sole Noise Barrier Construction Project / Retrofit Noise Barrier Construction.
When noise barrier construction is the sole or primary element of the construction project, then the costs of items, such as mobilization, insurance, and maintenance of traffic are attributable solely to the barrier construction. This situation, typically associated with a Type II or retrofit barrier construction, adds costs to the barrier construction, as compared to the cost of a similar barrier when part of a larger construction project. These additional costs can double the cost of the barrier installation. The equipment, materials, and manpower associated with such construction is required for only one task (noise barrier construction and its associated traffic control), with little opportunity for economy of scale applications.

13.2 Physical Conditions and Factors

13.2.1 Accessability. The ability to reach the noise barrier location will also influence the overall barrier cost. Barriers on top of large cut sections may require large cranes to lift barrier components (panels, posts, reinforcement, etc.) and construction equipment (drills, form work, etc.) to the site. In addition, concrete may need to be pumped to the site or even mixed on the site with materials lifted by crane. Barriers constructed in limited space areas (such as on structures, near traffic, etc.) will typically cost more to construct than barriers constructed in areas where freedom of movement is not restricted.

13.2.2 Transportation of Material, Equipment, and Work Force. In addition to the accessibility issues discussed above, the location of the barrier system components (earth, fabricated elements, etc.), the equipment (cranes, drills, form work, etc.) needed to erect the barrier, and the manpower to install the barrier are also factors influencing barrier cost. If barrier materials must be shipped from far away distances, costs will be greater than for locally obtained material. In addition, if specialized expertise is needed for a particular barrier installation, the required manpowers' proximity to the project will be a factor influencing cost.

13.2.3 Quantity of Barrier. The unit cost of a small quantity of a noise barrier will likely be greater than the unit cost for larger quantities of a barrier. The same or similar degree of insurance, mobilization, equipment type, and maintenance of traffic may be required to build 150 m (492 ft) of barrier as would be needed for 1500 m (4920 ft) of barrier. Form work, form liners, and similar fabrication layout is required if 50 or 500 noise panels are manufactured. Coating of a full roll of steel may be necessary, even if the full roll is not required for the particular project. Any specialized erection equipment or jigs will be required whether 10 or 100 barrier segments are erected.

13.2.4 Material Availability. Readily available materials will reduce barrier costs. If materials must be specially ordered, or if long manufacturing lead time is required, construction schedules can be affected, adding costs to the barrier construction. Any changes required by field adjustments, broken panels, etc., will be less costly if the material is readily available.

Further, since replacement of barrier elements may be required throughout the life of the noise barrier, the availability of replacement parts becomes a critical issue. To address this concern, some agencies have instituted a stock piling policy where the contractor/manufacturer, at the time of construction, supplies additional components to the organization responsible for maintenance. Typically, an additional 10 percent would by supplied to the responsible organization for stock piling purposes. The disadvantage to this type of practice is that the responsible organization, after several years of constructing noise barriers, may end up with an excessively large amount of varying stock on hand that has the potential of never being used. The issue of future availability becomes even more critical when the components have to be custom fitted with either very few or none of the pieces the same. In this situation, stock piling may not be an option. This consequence should be seriously considered during the design stage and should be avoided if at all possible.

13.2.5 Weather. Weather is a factor that can significantly affect barrier costs, particularly in areas where highway construction and noise barrier construction are restricted or limited during certain seasons. In these areas, fabrication schedules must be timed to have material available in non-restricted seasons. A delay in the fall delivery of noise barrier panels or posts would likely delay barrier construction until the spring. Unseasonable weather can either help or hinder barrier construction: an unusually mild winter may allow for otherwise restricted activities to continue, while an unseasonably cold winter may delay the thawing of ground necessary for certain construction activities. All of these factors will affect the ultimate cost of the noise barrier.

13.2.6 Traffic Protection and Detours. If traffic protection and/or traffic detours are a requirement, the cost of such protection may be attributed to the barrier cost. In addition, the contractor may charge a higher unit cost for barrier construction performed close to traffic as compared to construction in a less restricted area.

13.2.7 Limitation of Construction Hours.

- **Traffic Related** - Less efficient and more costly production can be expected when construction hours are limited or interrupted due to factors such as needing to keep lanes open during peak hours. Setup and knockdown time is costly and non-productive. If weekend or nighttime construction is restricted, it is possible that a project could take an additional construction season to complete.

- **Due to Local Noise Ordinances** - Similar types of inefficiencies (and associated increased costs) may occur in areas where local noise ordinances restrict the hours of construction.

- **Proximity to Neighborhoods** - Commitments made to communities to restrict either the hours of construction operations, the level of noise generated by such operations, or specific equipment used in the proximity of residential dwellings or other sensitive land uses will likely add cost to the noise barrier. Such increased costs can be related to items such as temporary noise barriers, quieter equipment, restricted operations, etc.

- **For Specific Operations - Pile Driving, Blasting, etc.** - Specific operations, such as pile driving, rock drilling, blasting, etc., may be restricted or prohibited on either a project-wide or specific area basis. They also may be limited to specific times and noise levels. Where such procedures are the only available technique, these limitations will undoubtedly result in increased barrier costs.

- **Interference with Utilities, Structures, and Other Features** - Increased noise barrier costs will result if construction requires the relocation of utilities or special barrier designs to avoid utilities. In addition, the construction of a barrier parallel to and underneath an overhead power line will require specialized equipment to erect the barrier or may prohibit the use of lift-in panels. Cast-in-place barriers may be required in these areas.

13.3 Labor Costs

Labor costs vary widely from area-to-area and thus have a significant bearing on overall noise barrier costs. In addition, certain barriers can be erected using general laborers, while other systems may require labor with specialized experience and expertise. Barriers constructed in areas where union trade agreements are in place may require numerous union trades to be involved in the barrier construction; while construction of a similar barrier in a location not governed by such agreements may be accomplished with general laborers.

Section Summary

Item #	Main Topic	Sub-Topic	Consideration	See Also Section	✓
13-1	Physical Conditions and Factors	Transportation of Material, Equipment, and Work Force	If barrier materials must be shipped from far away distances, costs will be greater than for locally obtained material. If specialized expertise is needed for a particular barrier installation, the required manpowers' proximity to the project will be a factor influencing cost.	13.1.2	
		Quantity of Barrier	The unit cost of a small quantity of a noise barrier will likely cost more than the unit cost for larger quantities of a barrier.	13.1.3	
		Material Availability	If materials must be specially ordered, or if long manufacturing lead time is required, construction schedules can be affected, adding costs to the barrier construction.	13.1.4	
		Traffic Protection and Detours	The cost of traffic protection/detours may increase barrier installation cost. The contractor may charge a higher unit cost for barrier construction performed close to traffic as compared to construction in a less restricted area.	13.1.6	
		Limitation of Construction Hours	Less efficient and more costly production can be expected when construction hours are limited or interrupted. Commitments made to communities to restrict either the hours of construction operations, the level of noise generated by such operations, or specific equipment used in the proximity of residential dwellings or other sensitive land uses will likely add cost to the noise barrier.	13.1.7	

14. BARRIER DESIGN PROCESS

This section describes several general topics involved in the barrier design process. Since each responsible organization has specific guidelines and/or policies related to the type and schedule of the various elements (acoustical, engineering, community involvement) of such a process, some aspects will only be briefly discussed here. State policies related to highway traffic noise are provided in the accompanying CD-ROM.

14.1 Acoustical Evaluation

The first step in the barrier design process is an acoustical evaluation. An acoustical evaluation is performed prior to the construction of a new highway or the expansion of an existing one to determine if noise abatement is needed and, if so, to what degree. General steps in performing an acoustical evaluation are as follows:

- Select noise sensitive receivers and/or areas for measurement and analysis (see Section 14.1.1);
- Determine **existing noise levels** by measurements and/or modeling (see Section 14.1.2);
- Determine if there will be any future noise impacts (see Section 14.1.3); and
- Determine the feasibility and reasonableness of noise abatement (see Section 14.1.4).

14.1.1 Select Noise Sensitive Receivers and/or Areas for Measurement and Analysis.
Site selection should be guided by the location of noise-sensitive receivers. Land-use maps and field reconnaissance should be used to identify potential noise-sensitive areas. For obvious reasons, schools, hospitals, and churches are especially sensitive to noise impacts. Noise sensitive residential areas should also be included in a noise-impact assessment. When selecting potential representative sites, keep in mind that the site should exhibit typical conditions (e.g., **ambient**, roadway infrastructure, and meteorological) for the surrounding area. It is recommended that good engineering judgment be used to select sites, keeping in mind the objectives of the study.[19]

14.1.2 Determine Existing Noise Levels by Measurements and/or Modeling.
Once the desired noise sensitive sites have been selected, the existing noise levels should be determined for comparison with estimated future noise levels (see Section 14.1.3). The results of this comparison, in concert with FHWA noise abatement criteria and with the responsible organization's policy, should be used to determine the appropriate noise abatement, if any (see Section 14.1.4). Existing noise levels at the desired sites are typically determined from either noise measurements (see Section 14.1.2.1) and/or noise modeling (see Section 14.1.2.2).

14.1.2.1 Noise Measurements.
This section describes briefly the recommended instrumentation, microphone location, sampling period, measurement procedures, and data analysis procedures to be used for performing roadside noise measurements. If measurement of noise levels are desired after a barrier is built (to determine a barrier's effectiveness), refer to Section 15. The procedures described below are in accordance with the FHWA's "Measurement of Highway-Related Noise."[19] Readers may refer to this document for more detailed discussions on all of the topics contained herein. Also included in this reference are sample field data log sheets.

Preceding Page Blank

- **Acoustic Instrumentation** - Figure 255 presents a generic, acoustic measurement instrumentation setup. All acoustic instrumentation should be calibrated annually by its manufacturer or other certified laboratory to verify accuracy. Where applicable, all calibrations shall be traceable to the National Institute of Standards and Technology (NIST).

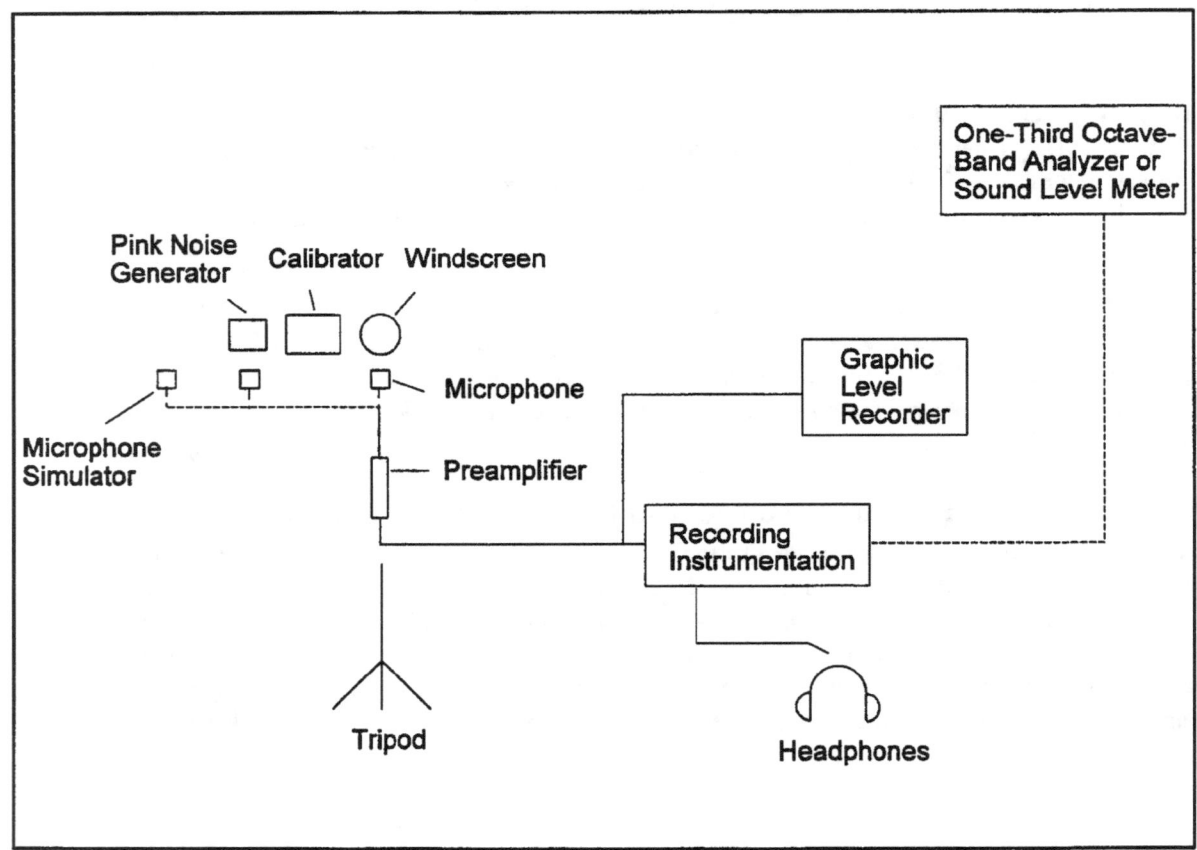

Figure 255. Generic measurement instrumentation setup.

- *Calibrator* - An acoustic calibrator provides a means of checking the entire acoustic instrumentation system's (i.e., microphone, cables, and recording instrumentation) sensitivity by producing a known sound pressure level (referred to as the calibrator's reference level) at a known frequency, typically 94 or 114 dB at 1 kHz or 124 dB at 250 Hz. The calibrator used for measurements described herein should meet the Type 1L performance requirements of ANSI S1.40-1984(R1997) or IEC 60942.[68,69]

- *Microphone Simulator* - The electronic noise floor of the entire acoustic instrumentation system should be established on a daily basis by substituting the measurement microphone with a passive microphone simulator (dummy microphone) and recording the noise floor for a period of at least 30 seconds. A dummy microphone electrically simulates the actual microphone by providing a known fixed (i.e., passive) capacitance which is equivalent to the minimum capacitance the microphone is capable of providing. This allows for valid measurement of the system's electronic noise floor.[70]

- *Pink Noise Generator* - The frequency response characteristics of the entire acoustic instrumentation system should be established on a daily basis by measuring and storing 30 seconds of pink noise. Pink noise is a random signal for which the **spectrum** density; i.e., narrow-band signal, varies as the inverse of frequency. In other words, one-third octave-band spectral analysis of pink noise yields a flat response across all frequency bands.

- *Windscreen* - A windscreen should be placed atop all microphones used in outdoor measurements. A windscreen is a porous sphere placed atop a microphone to reduce the effects of wind-generated noise on the microphone diaphragm. The windscreen should be clean, dry, and in good condition (a new windscreen is preferred). Typically, the effect on the measured sound level due to the insertion of a windscreen into an acoustic instrumentation system can be neglected.

- *Microphone System (Microphone and Preamplifier)* - A microphone transforms sound pressure variations into electrical signals that are, in turn, measured by instrumentation such as a sound level meter, a one-third octave-band spectrum analyzer, or a graphic level recorder. These electrical signals are also often recorded on tape for later off-line analysis. Microphone characteristics are further addressed in ANSI S1.4-1983(R1997).[12] A compatible preamplifier, if not engineered as part of the microphone system, should also always be used. A preamplifier provides high-input impedance and constant, low-noise amplification over a wide frequency range. Also, depending upon the type of microphone being used (refer to Reference 18), a preamplifier may also provide a polarization voltage to the microphone.

  The microphone system (microphone and preamplifier) should be supported using a tripod or similar device, such as an anchored conduit. Care should be taken to isolate the microphone system from the support, especially if the support is made up of a metal composite. In certain environments, the support can act as an antenna, picking up errant radio frequency interference which can potentially contaminate data. Common isolation methods include encapsulating the microphone system in nonconductive material (e.g., nylon) prior to fastening it to the support.

- *Graphic Level Recorder* - A graphic level recorder (GLR) connected to the analog output of the measuring or recording instrumentation is typically used in the field to provide a visual, real-time history of the measured noise level. A GLR plot varies in level at a known, constant pen-speed rate and response time that may be adjusted to approximate both slow and fast **exponential time-averaging**. It is valuable in visually judging ambient levels and verifying the acoustic integrity of individual events in real-time during measurements.

- *Recording Instrumentation* - There are two basic types of tape recorders: analog and digital. Analog recorders store signals as continuous variations in the magnetic state of the particles on the tape. Digital recorders store signals as a combination of binary "1s" and "0s." Not all field measurement systems will include a tape recorder. A recorder offers the unique capability of repeated playback of the measured noise source, thus allowing for more detailed analyses. The electrical characteristics of a tape recorder shall conform to the guidelines set in IEC 1265 and ANSI S1.13-1995 for frequency response and signal-to-noise ratio.[71,72]

- *One-Third Octave-Band Analyzer* - When the frequency characteristics of the sound source being measured are of concern, a one-third octave-band analyzer should be employed. In most cases, such a unit would not be employed directly in the field but would be used subsequent to field measurements in tandem with tape-recorded data. Such units can be employed to determine noise spectra, as well as compute various noise descriptors (see Section 3.2). Use of octave-band analyzers is not precluded; however, one-third octave-band analyzers are preferred.

- *Sound Level Meter* - For the purposes of all measurements discussed herein, sound level meters (SLMs) should perform true numeric integration and averaging in accordance with ANSI S1.4-1983(R1997).[12] Selection of a specific model of sound level meter should be based upon cost and the level of measurement accuracy desired.

- **Meteorological Instrumentation** - It is recommended that meteorological data, including temperature, relative humidity, and wind speed and direction be measured simultaneously with all acoustic data. For microphone distances within 30 m (100 ft) of the noise source, atmospheric effects, especially air turbulence, can affect measured sound levels (see Section 3.3.3). These effects typically increase with increasing distance from the noise source. Meteorological equipment shall include:

  - *Anemometer* - An anemometer is an instrument used to measure wind speed. Anemometers shall meet the requirements of ANSI S12.18-1994.[10] For general-purpose measurements at distances within 30 m from the source, a hand-held, wind-cup anemometer and an empirically observed estimation of wind direction are sufficient to document wind conditions. For research purposes, or for measurements where the receiver(s) will be positioned at distances greater than 30 m, a high-precision anemometer, capable of measuring wind conditions in three dimensions, integrated into an automated, data-logging weather station, should be used. For all types of measurements, the anemometer should be located at a relatively exposed position and at an elevation of at least 1.8 m (6 ft), preferably at the maximum height reached by the sound during propagation from source to receiver.[9]

  - *Thermometer, Hygrometer, and Psychrometer* - A thermometer for measuring ambient temperature and a hygrometer for measuring relative humidity should be used in conjunction with all noise measurement studies. An alternative is to use a psychrometer which is capable of measuring both dry and wet bulb temperature. Dry and wet bulb temperatures can then be used to compute relative humidity. For general purpose measurements, use of a sling psychrometer is recommended. For research purposes, a high-precision system may be needed, such as an automated, fast-response, data-logging weather station.

- **Traffic Instrumentation** - For many transportation-related measurements, the collection of traffic data, including the logging of vehicle types, vehicle-type volumes, and average vehicle speed may be required for: (1) determination of site equivalence (see Section 15.1.2); or (2) input into a highway traffic noise prediction model. This section discusses various instruments for the counting and classification of roadway traffic, including the use of a video camera, counting board, or pneumatic line. If none of these instruments is available, meticulous pencil/paper tabulation should be used.

- *Video Camera Recorder* - A video camera can be used to record traffic in the field and perform counts off-line at a later time. This approach, however, would require strict time synchronization between the acoustic instrumentation and the camera.

- *Counting Board* - A counting board is simply a board with three or more incrementing devices, depending on the number of vehicle types. Each device is manually triggered to increment for a given type of vehicle pass-by.

- *Pneumatic Line* - A pneumatic line may also be used to determine traffic counts. The pressure in the line increases when a vehicle passes over it, causing a mechanical switch to close. The mechanical switch triggers an internal counting mechanism to increment. The disadvantage of using a pneumatic line is that the specific vehicle mix; i.e., automobiles versus trucks, as well as other vehicle types, is not preserved.

- **Microphone Location**

  - *Reference Microphone* - The use of a reference microphone is strongly recommended for all existing-noise measurements. Use of a reference microphone allows for the application of adjustments to account for temporal variations in the noise source; e.g., traffic speeds, volumes, and mixes.

    Typically, the reference microphone is positioned at a height of 1.5 m (5 ft) above local ground level and located within 30 m (100 ft) of the centerline of the near travel lane at a position which is minimally influenced by ground and atmospheric effects. In addition, site geometry may dictate other reference microphone locations.

  - *Receiver Positions* - In most situations, study objectives will dictate specific microphone locations. Sometimes a single, typical residential area near the existing or proposed highway route can be used to represent other similar areas. If traffic conditions or topography vary greatly from one residential area to the next, receivers at many locations may be required.

    Receivers are also typically positioned at a height of 1.5 m (5 ft) above local ground level. However, microphone height(s) should be chosen to represent noise-sensitive receivers; i.e., if multi-story structures are of interest, including microphones at heights of 4.5 m and 7.5 m (15 ft and 25 ft) may be helpful. Microphone heights should be chosen to encompass the range of heights associated with all noise-sensitive receivers of interest.

- **Sampling Period** - Different sound sources require different sampling periods. Depending upon the characteristics of the sound source, a longer sampling period is needed to obtain a representative sample, averaged over all conditions. Typical sampling periods range from 2 to 30 minutes. In special instances where the temporal variation is expected to be substantial, longer sampling periods, such as 1 hr or 24 hr, may be necessary. Measurement repetitions at all receiver positions are required to ensure statistical reliability of measurement results. A minimum of 3 repetitions for like conditions is recommended, with 6 repetitions being preferred. Table 6 presents suggested measurement sampling periods based on the temporal nature and the range in sound level fluctuations for a particular sound source. Guidance on judgment of the temporal nature of the sound source may also be found in ANSI S1.13-1995 and ANSI S12.9-1988.[70,72]

Table 6. Sampling period.

Temporal nature	Greatest anticipated range		
	10 dB	10-30 dB	>30 dB
Steady *	2 minutes	N/A	N/A
Non-steady fluctuating	5 minutes	15 minutes	30 minutes
Non-steady intermittent	For at least 10 events	For at least 10 events	For at least 10 events
Non-steady, impulsive isolated bursts	For at least 10 events	For at least 10 events	For at least 10 events
Non-steady, impulsive-quasi-steady	3 cycles of on/off	3 cycles of on/off	3 cycles of on/off

* A minimum of three repetitions is recommended, with 6 repetitions being preferred.

■ Measurement Procedures

1. Prior to initial data collection, at hourly intervals thereafter, and at the end of the measurement day, the entire acoustic instrumentation system should be calibrated. Meteorological conditions (temperature, relative humidity, wind speed and direction, and cloud cover) should be documented prior to data collection, at a minimum of 15-minute intervals and whenever substantial changes in conditions are observed.

2. The electronic noise floor of the acoustic instrumentation system should be established daily by substituting the measurement microphone with a dummy microphone. The frequency response characteristics of the system should also be determined on a daily basis by measuring and storing 30 seconds of pink noise from a random-noise generator.

3. Ambient levels should be measured and/or recorded by sampling the sound level at each receiver and at the reference microphone, with the sound source quieted or removed from the site. A minimum of 10 seconds should be sampled. Note: If the study sound source cannot be quieted or removed, an upper limit to the ambient level using a statistical descriptor, such as L_{10}, may be used. Such upper limit ambient levels should be reported as "assumed." Note: Most sound level meters have the built-in capability to determine this descriptor.

4. Sound levels should be measured and/or recorded simultaneously with the collection of traffic data, including the logging of vehicle types, vehicle-type volumes, and the average vehicle speed. It is often easier to videotape traffic in the field and perform counts at a later time. This approach, of course, requires strict time synchronization between the acoustic instrumentation and the video camera. The videotape approach can also be used to determine vehicle speed.

- **Data Analysis Procedures**
    1. Adjust measured levels for calibration drift as follows:

        If the final calibration of the acoustic instrumentation differs from the initial calibration by greater than 1 dB, all data measured with that system during the time between calibrations should be discarded and repeated; and the instrumentation should be thoroughly checked.

        If the final calibration of the acoustic instrumentation differs from the initial calibration by 1 dB or less, all data measured with that system during the time between calibrations should be adjusted by arithmetically adding to the data the following CAL adjustment:

        $$\text{CAL adjustment} = \text{reference level} - [(\text{CAL}_{INITIAL} + \text{CAL}_{FINAL}) / 2] \quad (dB)$$

        For example:
        - reference level = 114.0 dB
        - initial calibration level = 114.1 dB
        - final calibration level = 114.3 dB

        Therefore:
        CAL adjustment = 114.0 - [(114.1+114.3)/2] = -0.2 dB

    2. Adjust measured levels for ambient as follows:

        If measured levels do not exceed ambient levels by 4 dB or more; i.e., they are masked, or if the levels at the reference microphone do not exceed those at the receivers, then those data should be omitted from analysis.

        If measured levels exceed the ambient levels by between 4 and 10 dB, and if the levels at the reference microphone exceed those at the receivers, then correct the measured levels for ambient as follows (Note: For source levels which exceed ambient levels by greater than 10 dB, the ambient contribution becomes essentially negligible and no correction is necessary):

        $$L_{adj} = 10 * \log_{10}(10^{0.1 L_c} - 10^{0.1 L_a}) \quad (dB)$$

        where: L_{adj} is the ambient-adjusted measured level;
        L_c is the measured level with source and ambient combined; and
        L_a is the ambient level alone.

        For example:
        - L_c = 55.0 dB
        - L_a = 47.0 dB

        Therefore:
        L_{adj} = $10 * \log_{10}(10^{(0.1*55.0)} - 10^{(0.1*47.0)})$ = 54.3 dB

    3. Compute the mean sound level for each receiver by arithmetically averaging the levels from individual sampling periods.

14.1.2.2 Noise Modeling.
There are many noise prediction methodologies being used by the highway noise community.[29,73,74,75] The current state-of-the-art in highway traffic noise prediction is the FHWA Traffic Noise Model, Version 1.0 (FHWA TNM®). Readers are directed to TNM's Trainer CD-ROM, which provides a detailed tutorial on using TNM and three companion reports (TNM's User's Guide, Technical Manual, and data base report).[4,5,6,7]

Following is a list of site characteristics to be included in the modeled analysis. These site characteristics can be determined from site visits, photos, aerial plans, etc.

- Roadways: coordinates, including roadway shoulder, vehicle types, traffic counts, vehicle speeds, and interrupted-flow devices, such as stop signs, traffic signals, etc.;
- Receiver: coordinates and height above ground;
- Existing noise barriers or barrier-like objects: barrier type (wall or berm), coordinates, height above ground, and absorptive characteristics;
- Building rows: coordinates, height above ground, and building percentage (the percentage of actual building structure in a row of buildings);
- Ground zones: coordinates and ground zone acoustic characteristics; and
- Terrain lines: coordinates which define substantial changes in ground elevation.

14.1.3 Determine If There Are Any Future Noise Impacts.
Existing noise levels are compared to future noise levels to establish if there will be any future noise impacts on the surrounding area. A noise impact associated with highway traffic noise occurs when: (1) predicted future noise levels exceed **existing levels** by a State Highway Agency (SHA) determined amount (i.e., "substantial increase"); and/or (2) predicted future noise levels approach or exceed the SHA's **impact criterion level**. This level is typically at least 1 dB(A) less than FHWA's Noise Abatement Criterion.

General steps in determining the need for noise abatement are as follows:[76]

- Determine future noise levels for all applicable alternatives - A noise prediction methodology (see Section 14.1.2.2) should be used to predict future noise levels for two cases:

  - *No build* - The predicted future noise levels without the planned highway project; and

  - *Build* - The predicted future noise levels after completion of the planned highway project, but without any noise abatement;

- Compare the predicted noise levels for all project alternatives (including the No-Build case) with FHWA noise abatement criteria and existing noise levels to see if there are any noise impacts as defined by the FHWA and SHA;

- If noise impacts have been identified, consider noise abatement measures. Noise abatement measures may include, but not limited to:

  1. Avoiding project impact by using design alternatives that result in a reduction in the noise effect, such as altering horizontal and vertical alignments;

2. Using traffic management measures that are consistent with State statutes regarding the regulation of traffic control devices, vehicle types, time-use restrictions, modified speed limits, etc.;
3. Acquiring property of interest to serve as a buffer zone to preempt development which would be impacted by traffic noise;
4. Constructing noise barriers (sound walls or earth berms); and
5. Insulating and/or air-conditioning public use or nonprofit institutional structures.

14.1.4 Determine Feasibility and Reasonableness of Noise Abatement.

The feasibility and reasonableness of noise abatement measures are based on the responsible organization's established criteria. A noise abatement measure is primarily considered feasible if the minimum **noise reduction goal** (as defined by the SHA) can be achieved at impacted receivers. However, keep in mind that even if an acoustically effective barrier may be feasible in theory (from computer modeling of the site), there may not be a physical way to actually build it making it unfeasible.

The determination of reasonableness is more subjective. Some factors typically considered in the determination of reasonableness include, but are not limited to:

1. Cost of abatement;
2. Amount of noise impact;
3. Noise abatement benefits;
4. Life cycle of abatement measures;
5. Environmental impacts of abatement measure;
6. Views of impacted residents; and
7. Input from public and local organizations.

14.2 Develop Barrier Design

If the acoustical evaluation has determined that a noise barrier is appropriate and both feasible and reasonable, the barrier design process begins. General steps in developing the barrier design are as follows:

- Using the input from the acoustical evaluation, develop the plan, profile, and cross sections of one or several barrier acoustical scenarios, locations, heights, lengths, and estimated costs;

- Document details of the recommended barrier(s) and transmit them to personnel responsible for designing the noise barrier;

- Have the design reviewed by staff who performed acoustical evaluation to ensure that any modifications based on engineering requirements did not adversely affect the acoustical performance of the barrier;

- Refine the design, as appropriate;

- Develop accompanying specifications, special provisions, etc.; have appropriate sections reviewed by acoustical personnel; and

- Develop the final Plans, Specifications, and Estimate (PS&E) package.

It is important to note that the steps listed above should be considered part of an iterative process where both the acoustical and civil engineers work together to resolve conflicts which arise during the design of a noise barrier. A balance between conflicting issues must be the ultimate goal, while preserving the primary function of the barrier system, which is effective and substantial noise reduction.

14.2.1 Community Participation.

Most agencies have specific guidelines and/or regulations which dictate the process of involving the public in the design of a noise barrier. The type, manner, and timing of community involvement activities vary widely between agencies. Many State policies are included on the CD-ROM which accompanies this Handbook. In general, community involvement programs normally include consideration of the following:

- Process options and techniques: formal and informal;
- Public meeting/hearing issues; and
    - Schedule, number, and progression of meetings and/or field views;
    - Location of meetings: in residences, churches, schools, public meeting places, or in the field;
    - Group size and participants;
    - Display materials: samples, photos, charts, plans, noise panels, etc.; and
    - Acoustical demonstration techniques: FHWA videos and project specific noise tapes.[17,26,77,78,79,80]
- Barrier system design options;
    - How presented;
    - What choices are provided to the public; and
    - Who is the public.

15. ASSESSING BARRIER EFFECTIVENESS

Following the construction of a noise barrier system, it may be necessary to evaluate the barrier's acoustical and non-acoustical performance. Such an evaluation may be required for several reasons, including commitments made during the design and/or environmental analysis phases of the project's development or the need to respond to comments related to the barrier's effectiveness made by community residents and motorists.

15.1 Acoustic Effectiveness

By comparing the barrier's insertion loss with the design goal insertion loss, the effectiveness of a barrier after construction can be assessed. The insertion loss (IL) is the sound level at a given receiver before the construction of a barrier minus the sound level at the same receiver after the construction of the barrier. IL may also be expressed as the net effect of barrier diffraction, combined with a partial loss of soft-ground attenuation. The partial loss of soft-ground attenuation is due to the barrier forcing the sound to take a higher path relative to the ground plane. To assess the effectiveness of a barrier system, the following basic steps should be followed:

- Select noise sensitive receivers and/or areas for measurement and analysis (see Section 15.1.1);
- Determine barrier insertion loss by measurements and/or modeling (see Section 15.1.2); and
- Compare measured/predicted insertion loss with the design goals of the barrier project.

15.1.1 Select Noise Sensitive Receivers and/or Areas for Measurement and Analysis.
Site selection should be guided by the location of noise-sensitive receivers. Land-use maps and field reconnaissance should be used to identify potential noise-sensitive areas. For obvious reasons, schools, hospitals, and churches are especially sensitive to noise impacts. Noise sensitive residential areas should also be included in a noise-impact assessment. When selecting potential representative sites, note that the site should exhibit typical conditions (e.g., ambient, roadway infrastructure, and meteorological) for the surrounding area. It is recommended that good engineering judgment be used to select sites, keeping in mind the objectives of the study.[19]

15.1.2 Determine Barrier Insertion Loss by Measurements and/or Modeling.
The procedures described in this section are in accordance with the American National Standards Institute ANSI S12.8-1998, "Methods for Determining the Insertion Loss of Outdoor Noise Barriers,"[9] which provides three methods to determine the field insertion loss of noise barriers: (1) the direct measured method; (2) the indirect measured method; and (3) the indirect predicted method. Readers may also refer to the FHWA's "Measurement of Highway-Related Noise"[19] for more detailed discussions on all of the topics contained herein. Also included in this reference are sample field data log sheets.

1. **The direct measured method** (see Section 15.1.2.1) may be used only if the barrier has not yet been installed or can be removed. Measurements are performed without the barrier to determine "BEFORE" sound levels, and another set of measurements are performed at the same site after construction to determine "AFTER" sound levels.

Note: For a valid determination of barrier insertion loss, BEFORE and AFTER measurements should be made under equivalent source, site, and atmospheric conditions, defined as follows:

- Equivalence in source conditions includes the number and mix of roadway traffic, as well as spectral content, directivity, spatial and temporal patterns, vertical and horizontal positions, and operating conditions of the individual vehicles. To a certain degree, non-equivalence in traffic conditions can be factored out through the use of a reference microphone (See Section 15.1.2.1).

- Equivalence in site geometry entails similar terrain characteristics and **ground impedance** within an angular sector of 120 degrees from all receivers looking toward the noise source. For research purposes, equivalence in ground impedance may be determined by performing measurements in accordance with the ANSI Standard for measuring ground impedance scheduled for publication in 1999.[13] For more empirical studies, or if measurements are not feasible, then the ground for BEFORE and AFTER measurements may be judged equivalent if general ground surface type and conditions; e.g., surface water content, are similar.

- Equivalence in meteorological conditions includes wind, temperature, humidity, and cloud cover. Wind conditions may be judged equivalent for BEFORE and AFTER measurements if the **wind class** (see Table 7) remains unchanged and the vector components of the average wind velocity from source to receiver do not differ by more than a certain limit, which is defined as follows: (1) for an acoustical error within ±1.0 dB and distances less than 70 m (230 ft), this limit is 1.0 m/s (2 mi/h); (2) for an acoustical error within ±0.5 dB and distances less than 70 m (230 ft), at least four BEFORE and AFTER measurements should be made within the limit of 1.0 m/s (2 mi/h). However, these 1.0 m/s limits are not applicable for a calm wind class when strong winds with a small vector component in the direction of propagation exist. In other words, BEFORE/AFTER measurements in such instances should be avoided.

Table 7. Classes of wind conditions.

Wind class	Vector component of wind velocity (m/s)
Upwind	-1 to -5
Calm	-1 to +1
Downwind	+1 to +5

Average temperatures during BEFORE and AFTER measurements may be judged equivalent if they are within 14 degrees Celsius of each other. Also, in certain conditions, dry air produces substantial changes in the atmospheric absorption of sound at high frequencies. Therefore, for a predominantly high-frequency source (most sound energy over 3000 Hz), the absolute humidity for BEFORE and AFTER measurements should be similar; e.g., within 20 percent.

The BEFORE and AFTER acoustical measurements should be made under the same class of cloud cover (See Table 8).

Table 8. Classes of cloud cover.

Class	Description
1	Heavily overcast
2	Lightly overcast (either with continuous sun or the sun obscured intermittently by clouds 20 to 80 percent of the time)
3	Sunny (sun essentially unobscured by clouds 80 percent of the time)
4	Clear night (less than 50 percent cloud cover)
5	Overcast night (50 percent or more cloud cover)

The advantage of using this method is that it ensures identical site geometric characteristics. However, the disadvantages are that equivalent meteorological and traffic conditions may not be reproducible.

2. **The indirect measured method** (see Section 15.1.2.1) may be used when the barrier has been installed prior to any direct BEFORE measurements and cannot be removed to permit such measurements. In this case, the BEFORE condition is simulated at an equivalent site without the barrier. In this case, BEFORE and AFTER measurements should be performed simultaneously at adjacent locations, if possible. The primary advantage of using this method is that it ensures essentially the same meteorological and traffic conditions. The difficulty is that an adjacent equivalent site may not always be available. If an adjacent equivalent site is available, then this method is preferred over the direct measured method.

Note: For a valid determination of barrier insertion loss, BEFORE and AFTER measurements should be made under equivalent source, site, and atmospheric conditions as discussed for the direct measured method.

3. **The indirect predicted method** may be used if neither direct BEFORE measurements or indirect BEFORE measurements at an equivalent site are possible. In this case, BEFORE levels are predicted using a highway-traffic, noise-prediction model, such as the Federal Highway Administration's Traffic Noise Model (FHWA TNM®).[4,5,6,7] The resulting insertion loss should be referred to as "partially measured." This method is inherently the least accurate of the three methods presented herein (see Section 15.1.2.2).

15.1.2.1 Noise Measurements.
This section describes briefly the recommended instrumentation, microphone location sampling period, measurement procedures, and data analysis procedures to be used for performing BEFORE and/or AFTER measurements using any of the three methods described above.

- **Instrumentation** - Figure 255 in Section 14.1.2.1 presents and describes a generic, acoustic measurement instrumentation setup. All acoustic instrumentation should be calibrated annually by its manufacturer or other certified laboratory to verify accuracy. Where applicable, all calibrations shall be traceable to the National Institute of Standards and Technology (NIST).

- **Microphone Location** - When performing measurements to determine barrier insertion loss, it is important to remember that microphone locations relative to the sound source in the BEFORE and AFTER cases should be as close to identical as possible.

  - *Reference Microphone* - The use of a reference microphone is strongly recommended. Use of a reference microphone allows for a calibration of measured levels, which accounts for variations in the characteristics of the noise source; e.g., traffic speeds, volumes, and mixes.

    In most cases, a reference microphone is placed between the noise source and other measurement microphones at a height of at least 1.5 m (5 ft) directly above the top edge of the barrier (see Figure 256), and at a distance from the sound source sufficient to minimize near-field effects. Typically, a minimum, standard distance of 15 m (50 ft) from the noise source is used. If the barrier is located less than 15 m from the source, the reference microphone should be placed at a distance of 15 m from the noise source, but at a height such that the line of sight between the microphone and the ground plane beneath the source is at least 10 degrees (see Figure 257). This location should remain the same for all measurements, including measurements at the equivalent site, where the barrier is not present.*

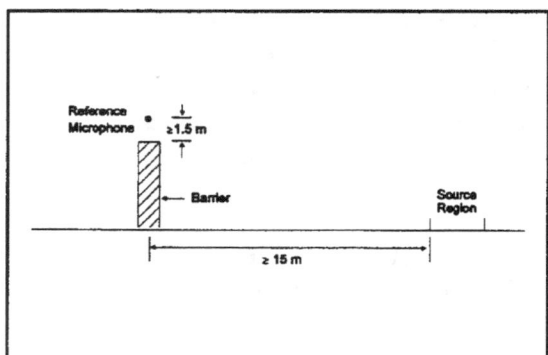

Figure 256. Reference microphone - position 1.

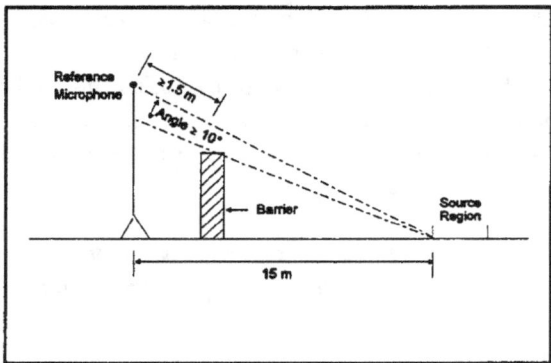

Figure 257. Reference microphone - position 2.

  - *Receiver Positions* - In most situations, study objectives will dictate specific microphone locations. As such, this section presents a very generic discussion of microphone locations and assumes no specific study objectives have been identified.

    Generally, it is convenient to position microphones at offset distances from the barrier which correspond to incremental doublings of distances (e.g., 15, 30, and 60 m [50, 100, and 200 ft]). Often times measurement sites are characterized by drop-off rates as a function of distance doubling.

---

* The reference microphone positions shown in Figures 256 and 257 are preferred. If the barrier is located more than 30 m from the source, and hence, probably closer to the receiver positions, locate the reference microphone at a distance of 15 m from the source and at a height such that a line from the near edge of the source to the microphone makes an angle greater than 10 degrees with the nominal ground plane.

In terms of microphone height, 1.5 m (5 ft) is the preferred position. If multi-story structures are of interest, including microphones at heights of 4.5 m and 7.5 m (15 ft and 25 ft) may be helpful. Microphone heights should be chosen to encompass all noise-sensitive receivers of interest (see Figure 258).

For the purpose of determining barrier insertion loss, it is important to remember that microphone locations relative to the sound source in the BEFORE and AFTER cases must be identical. There may be instances when receivers are placed on the lawns of homes within the community adjacent to a noise barrier.

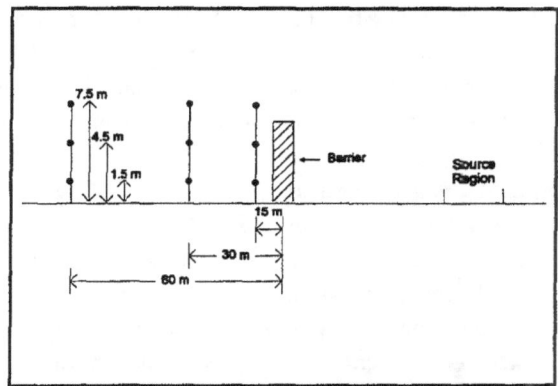

Figure 258. Receiver positions.

- **Sampling Period** - Different sound sources require different sampling periods. For multiple-source conditions, a longer sampling period is needed to obtain a representative sample, averaged over all conditions. Typical sampling periods range from 2 to 30 minutes. In special instances where the temporal variations are expected to be substantial, longer sampling periods, such as 1 hr or 24 hr, may be necessary. Measurement repetitions at all receiver positions are required to ensure statistical reliability of measurement results. A minimum of 3 repetitions for like conditions is recommended, with 6 repetitions being preferred. Table 7 in Section 14.1.2.1 presents suggested measurement sampling periods based on the temporal nature and the range in sound level fluctuations of the noise source. Guidance on judgment of the temporal nature of the source may also be found in ANSI S1.13-1995 and ANSI S12.9-1988.[70,72]

- **Measurement Procedures**
    1. Prior to initial data collection, at hourly intervals thereafter, and at the end of the measurement day, the entire acoustic instrumentation system should be calibrated. Meteorological conditions (temperature, relative humidity, wind speed and direction, and cloud cover) should be documented prior to data collection, at a minimum of 15-minute intervals and whenever substantial changes in conditions are observed.

    2. The electronic noise floor of the acoustic instrumentation system should be established daily by substituting the measurement microphone with a dummy microphone. The frequency response characteristics of the system should also be determined on a daily basis by measuring and storing 30 seconds of pink noise from a random-noise generator.

    3. Ambient levels should be measured and/or recorded by sampling the sound level at each receiver and at the reference microphone, with the sound source quieted or removed from the site. A minimum of 10 seconds should be sampled. Note: If the study sound source cannot be quieted or removed, an upper limit to the ambient level using a statistical descriptor, such as L_{10}, may be used. Such upper limit ambient levels should be reported as "assumed." Note: Most sound level meters have the built-in capability to determine this descriptor.

4. Sound levels should be measured and/or recorded simultaneously with the collection of traffic data, including the logging of vehicle types, vehicle-type volumes, and the average vehicle speed. It is often easier to videotape traffic in the field and perform counts at a later time. This approach, of course, requires strict time synchronization between the acoustic instrumentation and the video camera. The videotape approach can also be used to determine vehicle speed.

- **Data Analysis Procedures**
    1. For valid comparisons of BEFORE and AFTER measured levels, the equivalence of meteorological conditions; i.e., wind, temperature, humidity, and cloud cover, should be established. It is assumed that equivalence of site parameters, such as terrain characteristics and ground impedance, were established prior to performing measurements. Sampling periods in which equivalence cannot be established should be excluded from subsequent analysis.

    2. Adjust measured levels for calibration drift as follows:

        If the final calibration of the acoustic instrumentation differs from the initial calibration by greater than 1 dB, all data measured with that system during the time between calibrations should be discarded and repeated; and the instrumentation should be thoroughly checked.

        If the final calibration of the acoustic instrumentation differs from the initial calibration by 1 dB or less, all data measured with that system during the time between calibrations should be adjusted by arithmetically adding to the data the following CAL adjustment:

        $$\text{CAL adjustment} = \text{reference level} - [(\text{CAL}_{INITIAL} + \text{CAL}_{FINAL}) / 2] \qquad (\text{dB})$$

        For example:
        - reference level   = 114.0 dB
        - initial calibration level = 114.1 dB
        - final calibration level   = 114.3 dB

        Therefore:
        CAL adjustment = 114.0-[(114.1+114.3)/2] = -0.2 dB

    3. Adjust measured levels for ambient as follows:

        If measured levels do not exceed ambient levels by 4 dB or more; i.e., they are masked, or if the levels at the reference microphone do not exceed those at the receivers, then those data should be omitted from analysis.

        If measured levels exceed the ambient levels by between 4 and 10 dB, and if the levels at the reference microphone exceed those at the receivers, then correct the measured levels for ambient as follows (Note: For source levels which exceed ambient levels by greater than 10 dB, the ambient contribution becomes essentially negligible and no correction is necessary):

$$L_{adj}=10*\log_{10}(10^{0.1L_c} - 10^{0.1L_a}) \quad \text{(dB)}$$

where: L_{adj} is the ambient-adjusted measured level;
L_c is the measured level with source and ambient combined; and
L_a is the ambient level alone.

For example:
- L_c = 55.0 dB
- L_a = 47.0 dB

Therefore:
L_{adj} = $10*\log_{10}(10^{(0.1*55.0)}-10^{(0.1*47.0)})$ = 54.3 dB

4. If appropriate, determine the reflection and/or edge-diffraction bias adjustment.[*]

5. Compute the barrier insertion loss or lower-bound to insertion loss for each source-receiver pair as follows:

For each measurement repetition and each BEFORE/AFTER pair, the insertion loss, or its **lower bound**, should be determined by subtracting the difference in adjusted reference and receiver levels for the BEFORE case from the difference in adjusted reference and receiver levels for the AFTER case:

$$IL_i = (L_{Aref} + L_{edge} - L_{Arec}) - (L_{Bref} - L_{brec}) \quad \text{(dB)}$$

where: IL_i is the insertion loss at the i_{th} receiver;
L_{Bref} and L_{Aref} are, respectively, the BEFORE and AFTER adjusted reference levels;
L_{edge} is the reflection and/or edge-diffraction bias adjustment; and
L_{Brec} and L_{Arec} are, respectively, the BEFORE and AFTER adjusted source levels at the ith receiver.

For example:
- L_{Aref} = 78.2 dB
- L_{edge} = - 0.5 dB
- L_{Arec} at receiver 1 = 56.3 dB
- L_{Bref} = 77.7 dB
- L_{Brec} at receiver 1 = 65.0 dB

---

[*] Due to multiple reflections between source and barrier and/or edge diffraction at the top of a barrier, a 0.5 dB correction factor to reference microphone sound levels in the AFTER case may be applied. Good engineering judgment, based on repeatability through measurements, should be used to determine the magnitude and necessity of this correction. For example, if for several runs (i.e., greater than six), a consistent repeatable difference at the reference microphone position in the BEFORE and AFTER case occurs, and it can be proven that the traffic during both cases were equivalent, then the difference can be attributed to edge diffraction effects. The edge diffraction correction factor will be a negative value which is added directly to the sound level measured at the reference microphone in the AFTER case. Note: Larger corrections due to parallel barriers may be necessary.

Therefore:
$IL_1 = (78.2 - 0.5 - 56.2) - (77.7 - 65.0) = 21.4 - 12.7 = 8.7$ dB

Note: The lower bound to barrier insertion loss is the value reported when ambient levels are not directly measured without the sound source in operation; i.e., "assumed" ambient.

15.1.2.2 Noise Modeling.

As stated earlier, the indirect predicted method requires performing measurements at a site with a barrier to determine AFTER levels, and using a highway-traffic noise-prediction model to predict sound levels without a barrier. This method is inherently the least accurate of the three methods presented. There are many noise prediction methodologies being used by the highway noise community.[29,73,74,75] The current state-of-the-art in highway traffic noise prediction is the FHWA Traffic Noise Model, Version 1.0 (FHWA TNM®). Readers are directed to TNM's Trainer CD-ROM, which provides a detailed tutorial on using TNM, and to three companion reports (TNM's User's Guide, Technical Manual, and data base report).[4,5,6,7] This section describes briefly the procedures associated with the modeling approach.

1. Determine sound levels for the AFTER case according to Section 15.1.2.1.

2. Using the measured traffic data and the observed site data, input the necessary information into a highway-noise prediction model, such as the FHWA TNM to compute BEFORE levels at the reference position and at each receiver position. It is possible that modeled levels at the reference position may differ substantially in the BEFORE case, as compared with the measured AFTER case. In such instances, the difference observed at the reference microphone shall be used as a calibration factor for all other measurement positions.

   Following is a list of site characteristics to be included in the modeled analysis. These site characteristics can be determined from site visits, photos, aerial plans, etc.

   - Roadways: coordinates, including roadway shoulder, vehicle types, traffic counts, vehicle speeds, and interrupted-flow devices, such as stop signs, traffic signals, etc.;
   - Receiver: coordinates and height above ground;
   - Existing noise barriers or barrier-like objects: barrier type (wall or berm), coordinates, height above ground, and absorptive characteristics;
   - Building rows: coordinates, height above ground, and building percentage (the percentage of actual building structure in a row of buildings);
   - Ground zones: coordinates and ground zone acoustic characteristics; and
   - Terrain lines: coordinates which define substantial changes in ground elevation.

3. Compute the insertion loss according to Section 15.1.2.1. The resulting insertion loss should be referred to as "partially measured."

15.2 Non-acoustic Effectiveness

An evaluation of a barrier system's non-acoustic effectiveness can include objective factors such as structural durability of the barrier and the appearance of the barrier over time. These objective factors generally require that a considerable time period elapse following barrier construction before any evaluation can be made. On the other hand, subjective factors such as community acceptance of the noise barrier system and public perceptions can be evaluated in the time period shortly after the construction of the barrier, as long as the traffic volumes have stabilized at the facility. A wide range of sampling and interviewing techniques are available to obtain input from adjacent landowners and the motoring public.

15.2.1 Community Acceptance.

Opinions expressed by landowners immediately adjacent to a noise barrier typically reflect their perceptions of the noise levels actually heard versus those levels which they had expected to hear. For example, a barrier may be providing a full 10-dB reduction, but if the final with-barrier levels are still near 66 or 67 dBA, it still may be judged as "too loud" or "ineffective" or "no better than before." In addition, perceptions related to the barrier can relate to the intrusiveness of noise and/or the barrier from either an acoustical or non-acoustical standpoint. Landowners living farther from a noise barrier sometimes complain about noise generated by a new installation. In such situations, while noise levels are typically not loud enough to require consideration of noise abatement, a different or new noise exists as the result of a new or upgraded installation. Closer to the installation, residents adjacent to a highway along which a new barrier was constructed on the opposite side may complain about increased noise.

While the potential for opinions and perceptions such as those noted above is inevitable with the construction of any new barrier system, dealing with such factors is primarily a policy-related issue, as compared to a noise barrier design issue. Similar to the specific noise barrier design and analysis policies unique to each state, each state also has a specific and unique policy (whether formalized or not) related to community involvement, public participation, and post-construction evaluation factors. The reader is referred to the specific state noise policies (contained on the companion CD-ROM) for more information.

15.2.2 Cost.

All states have procedures for evaluating the cost and effectiveness of noise barrier systems. For the most part, a barrier's cost and acoustic effectiveness are factors considered in determining barrier feasibility and reasonableness. The reader is referred to the specific state noise policies (contained on the companion CD-ROM) for details related to the particular items evaluated in such determinations.

Since barrier feasibility and reasonableness evaluations are typically performed during the design phase, they are normally based on modeled noise levels and historical barrier costs. As such, they may not necessarily reflect the final and true cost or effectiveness of a specific as-constructed barrier. If a more refined cost-effectiveness value is desired, a post-construction evaluation may be performed to take into account the following post-construction conditions:

- Measured noise levels;
- Revised insertion loss calculations;
- Costs of modifications made during construction;
- Actual construction bid costs; and
- Actual material costs.

Evaluating costs in this detail may still not provide the true costs related to a specific barrier system, particularly on large projects for which the noise barriers comprise just one relatively small component. Front end loading of specific construction items is often found in construction bid documents for the purpose of establishing contract cash flow. On projects for which noise barrier construction is the major or only item, a determination must be made as to whether or not to include items such as insurance, maintenance, protection of traffic, mobilization, etc. in the barrier cost. More discussion on such cost implications is found in Section 13.

16. TOOLS TO ASSIST

This section discusses available tools to assist the highway engineer, barrier designer, and community participant in barrier design. These tools should be used to supplement the information described in this Handbook.

16.1 Barrier Design Video and CD-ROM

This Handbook is accompanied by a complementary videotape and CD-ROM.[2,3] The video highlights the important aspects of the Handbook using film footage and photographic examples of noise barrier designs. The CD-ROM contains the complete Handbook, the entire data base of photographs taken in support of this study (over 3000 total), State Transportation Agency (SHA) highway noise policies, as well as contact information. The photo data base is directly linked to related sections, subsections, and photographs within the Handbook itself. Section 16.5 also lists SHA personnel, their addresses, and phone numbers.

16.2 FHWA Traffic Noise Model

FHWA Traffic Noise Model, Version 1.0 (FHWA TNM) is an entirely new, state-of-the-art computer program used for predicting noise impacts in the vicinity of highways.[4,5,6,7] The FHWA TNM replaces the FHWA's prior prediction model and software, STAMINA 2.0/OPTIMA.[29,73] The FHWA TNM contains the following components:

- Modeling of five standard vehicle types, including automobiles, medium trucks, heavy trucks, buses, and motorcycles, as well as user-defined vehicles;

- Modeling of both constant-flow and interrupted-flow traffic using a 1994/1995 field-measured data base;

- Modeling of the effects of different pavement types, as well as the effects of graded roadways;

- Sound level computations based on a one-third octave-band data base and algorithms;

- Graphically-interactive noise barrier design and optimization;

- Attenuation over/through rows of buildings and dense vegetation;

- Multiple diffraction analysis;

- Parallel barrier analysis; and

- Contour analysis, including sound level contours, barrier insertion loss contours, and sound-level difference contours.

These components are supported by a scientifically founded and experimentally calibrated acoustic computation methodology, as well as an entirely new, and more flexible data base, as compared with that of its predecessor, STAMINA 2.0/OPTIMA. The data base is made up of over 6000 individual vehicle pass-by events measured at forty sites across the country. It is the primary building block around which the acoustic algorithms are structured.

Further information on TNM can be found on the McTrans Website (http://www-mctrans.ce.ufl.edu) or by contacting McTrans at:

        McTrans Center   PHONE: (352) 392-0378
        University of Florida   FAX: (352) 392-3224
        512 Weil Hall
        P.O. Box 116585
        Gainsville, FL 32611-6585

16.3 AASHTO, ANSI, ASTM, CSA, IEC, ISO, NCHRP, NIST, and SAE Standards

This Handbook cites as references many national and international standards. These standards contain technical specifications, methodologies, or other criteria to be used to ensure consistency in definitions, practices, materials, products, etc. The following provides contact information for pertinent standards organizations:

- The **American Association of State Highway and Transportation Officials (AASHTO)** is an advocate for multimodal and intermodal transportation providing technical services, information, and advice on transportation issues.

  American Association of State Highway and Transportation Officials
  444 North Capitol Street, NW, Suite 249
  Washington, DC 20001
  Telephone: (202) 624-5800
  Fax: (202) 624-5806
  http://www.aashto.org

- The **American National Standards Institute (ANSI)** was founded in 1918 to promote and facilitate voluntary consensus standards and conformity assessment systems.

  American National Standards Institute
  11 West 42nd Street
  New York, NY 10036
  Telephone: (212) 642-4900
  Fax: (212) 398-0023
  http://www.ansi.org

- The **American Society For Testing and Materials (ASTM)** develops and provides consensus standards, related technical information and services to promote public health and safety, and the reliability of materials, products, systems, and services to facilitate national, regional, and international commerce.

  American Society For Testing and Materials
  ASTM, 100 Barr Harbor Drive
  West Conshohocken, PA 19428-2959

FHWA Highway Noise Barrier Design Handbook — Tools to Assist

Telephone: (610) 832-9585
Fax: (610) 832-9555
http://www.astm.org

- The **Canadian Standards Association (CSA)** was chartered in 1919 as a non-profit association and has developed over 2000 standards covering the life sciences, environment, electrical and electronic products, communications systems, building construction, energy, transportation/distribution, materials technology, and quality business management.

  CSA International
  178 Rexdale Boulevard
  Etobicoke (Toronto), Ontario M9W 1R3
  Telephone: (800) 463-6727
  Fax: (416) 747-4149
  http://www.csa-international.org/english/home/index.html

- The **International Electrotechnical Commission (IEC)** was founded in 1906 as a result of a resolution passed at the International Electrical Congress held in St. Louis (USA) in 1904. IEC prepares and publishes international standards for all electrical, electronic, and related technologies.

  International Electrotechnical Commission
  3, rue de Varembé
  P.O. Box 131
  CH - 1211 Genève 20
  Switzerland
  Telephone: 41-22-919-02-11
  Fax: 41-22-919-03-00
  http://www.iec.ch

- The **International Organization for Standardization (ISO)** is a non-governmental organization established in 1947 to promote the development of standardization and related activities in the world with a view to facilitating the international exchange of goods and services, and to developing cooperation in the spheres of intellectual, scientific, technological, and economic activity. ISO's work results in international agreements which are published as International Standards.

  International Organization for Standardization
  1, rue de Varembé
  Case postale 56
  CH-1211 Genève 20
  Switzerland
  Telephone 41-22-749-01-11
  Fax   41-22-733-34-30
  http://www.iso.ch/index.html

- The **National Cooperative Highway Research Program (NCHRP)** was created in 1962 as a means to conduct research in acute problem areas that affect highway planning, design, construction, operation, and maintenance nationwide. NCHRP is administered by the Transportation Research Board (TRB) and sponsored by the member departments (i.e., individual state departments of transportation) of AASHTO, in cooperation with the FHWA.

National Cooperative Highway Research Program
Transportation Research Board
2101 Constitution Avenue NW
Washington, DC 20418
Telephone: (202) 334-2379
Fax: (202) 334-2006
http://www.aashto.org/prog_svcs/a_ps.html

- The **National Institute of Standards and Technology (NIST)** was established by Congress as an agency of the U.S. Department of Commerce's Technology Administration to assist industry in the development of technology needed to improve product quality, to modernize manufacturing processes, to ensure product reliability, and to work with industry to develop and apply technology, measurements, and standards.

National Institute of Standards and Technology
100 Bureau Drive
Gaithersburg, MD 20899-0001
Telephone: (301) 975-NIST (6478)
http://www.nist.gov

- The **Society of Automotive Engineers (SAE)** is a resource for technical information and expertise used in designing, building, maintaining, and operating self-propelled vehicles for use on land, sea, air, or space. The SAE publishes many new, revised, and reaffirmed standards each year in three categories: Ground Vehicle Standards (J-Reports); Aerospace Standards; and Aerospace Material Specifications (AMS).

SAE World Headquarters
400 Commonwealth Drive
Warrendale, PA 15096-0001
Telephone: (724) 776-4841
Fax: (724) 776-5760
http://www.sae.org

16.4 Technical Documents

The Reference section of this report cites many technical documents and field studies used in the development of this Handbook. The majority of those documents can be obtained either from the appropriate State Departments of Transportation (see Section 16.5) or from the National Technical Information Service (NTIS). NTIS is the federal government's central source for the sale of scientific, technical, engineering, and related business information produced by or for the U.S. government. NTIS also maintains complementary material from international sources.

National Technical Information Service
Technology Administration
U.S. Department of Commerce
Springfield, VA 22161
Telephone: 1-800-553-NTIS (6847) or (703) 605-6000
Fax: (703) 605-6900
http://www.ntis.gov

16.5 State Departments of Transportation

The following is a list of State Department of Transportation contacts for highway traffic noise:

Rick Mook
Environmental Technical Section
Alabama Dept. of Transportation
1409 Coliseum Blvd.
Montgomery, Alabama 36130
Telephone: 334-242-6576
Fax: 334-269-0826

Dave Bloom
Alaska Dept. of Transportation
2301 Peger Road
Fairbanks, Alaska 99701
Telephone: 907-451-2228
Fax: 907-451-5390

Liz Szews
Environmental Planning Services
Arizona Dept. of Transportation
205 South 17th Ave., Mail Drop 619E
Phoenix, Arizona 85007-3212
Telephone: 602-255-8642
Fax: 602-407-3066

Mike Webb
Arkansas Dept. of Transportation
P.O. Box 2261
Little Rock, Arkansas 72203
Telephone: 501-569-2281
Fax: 501-569-2009

Keith Jones
CALTRANS Environmental Program
1120 N Street, Mail Station 27
Sacramento, CA 94274-0001
Telephone: 916-653-2351
Fax: 916-653-7757

Makeba Adesunloye
Colorado Dept. of Transportation
Office of Environmental Review
4201 East Arkansas Ave., Room 284
Denver, Colorado 80222-3400
Telephone: 303-757-9016
Fax: 303-757-9445

Carmine Trotta
Connecticut Dept. of Transportation
P.O. Box 317546
2800 Berlin Turnpike
Newington, Connecticut 06131-7546
Telephone: 860-594-2939
Fax: 860-594-3028

Richard K. Vetter
Delaware Dept. of Transportation
P.O. Box 778
Dover, Delaware 19903
Telephone: 302-760-2134
Fax: 302-739-2251

Win Lindeman
Florida Dept. of Transportation
605 Suwannee Street, MS-37
Tallahassee, Florida 32399-0450
Telephone: 850-488-2914
Fax: 850-922-7217

Susan Knudson
Georgia Dept. of Transportation
3993 Aviation Circle
Atlanta, GA 30336
Telephone: 404-699-4407
Fax: 404-699-4440

Alfred Makino
Hawaii Dept. of Transportation
869 Punchbowl Street
Honolulu, Hawaii 96813
Telephone: 808-832-3557
Fax: 808-832-1787

Roy Jost
Environmental Planner
Idaho Dept. of Transportation
P.O. Box 7129
Boise, Idaho 83707-1129
Telephone: 208-334-8484
Fax: 208-334-3858

Mike Bruns
Noise Specialist
Illinois Dept. of Transportation
2300 South Dirksen Parkway, Rm 330
Springfield, Illinois 62764
Telephone: 217-782-7077
Fax: 217-524-9356

Steve Cecil
Indiana Dept. of Highways
Room 848, State Office Building
Indianapolis, Indiana 46204-2249
Telephone: 317-232-5468
Fax: 317-232-5478

Ron Ridnour
Office of Project Planning
Iowa Dept. of Transportation
Ames, Iowa 50010
Telephone: 515-239-1613
Fax: 515-239-1982

Scott P. Vogel
Environmental Services Section
Kansas Dept. of Transportation
Topeka, Kansas 66612
Telephone: 785-296-0853
Fax: 785-296-8399

Barry Adkins
Kentucky Transportation Cabinet
Division of Environmental Analysis
125 Holmes Street
Frankfort, Kentucky 40622-1994
Telephone: 502-564-7250
Fax: 502-564-5655

Noel Ardoin
Louisiana Dept. of Transp. & Development
P.O. Box 94245
Baton Rouge, Louisiana 70804-9245
Telephone: 225-929-9194
Fax: 225-929-9188

William S. Rollins
Maine Dept. of Transportation
16 State House Station
Augusta, Maine 04333-0016
Telephone: 207-287-3944
Fax: 207-287-6737

Ken Polcak
Maryland State Highway Adm.
Office of Environmental Design
707 N. Calvert Street C-305
Baltimore, MD 21202
Telephone: 410-545-8601
Fax: 410-333-4126

Mike Paiewonsky
Massachusetts Highway Dept.
10 Park Plaza, Room 4260
Boston, Massachusetts 02116-3973
Telephone: 617-973-8245
Fax: 617-973-8879

Tom Peek
Michigan Dept. of Transportation
P.O. Box 30050
Lansing, Michigan 48909
Telephone: 517-335-2616
Fax: 517-373-9255

Melvin Roseen
Minnesota Dept. of Transportation
Noise Analysis Unit
6000 Minnehaha Avenue, South
St. Paul, Minnesota 55111
Telephone: 612-725-2373
Fax: 612-725-2385

Elton D. Holloway
Planning Division
Mississippi Dept. of Transportation
P.O. Box 1850
Jackson, Mississippi 39215-1850
Telephone: 601-359-7685
Fax: 601-359-7652

Macey Jett
Missouri Highway & Transportation Dept.
P.O. Box 270
Jefferson City, Missouri 65102
Telephone: 573-526-5648
Fax: 573-526-3239

Cora G. Helm
Montana Dept. of Transportation
P.O. Box 201001
Helena, Montana 59620-1001
Telephone: 406-444-7659
Fax: 406-444-7245

Cynthia Veys
Project Development Division
Nebraska Department of Roads
P.O. Box 94759
Lincoln, Nebraska 68509-4759
Telephone: 402-479-4410
Fax: 402-479-4325

Mike Painter
Environmental Services Division
Nevada Dept. of Transportation
1263 South Stewart Street
Carson City, Nevada 89701-5229
Telephone: 702-888-7685
Fax: 702-888-7104

Charles Hood
Bureau of Environment Room 109
New Hampshire Dept. of Transportation
P.O. Box 483
Concord, New Hampshire 03302-0483
Telephone: 603-271-3226
Fax: 603-271-3914

Domenick Billera
New Jersey Dept. of Transportation
1035 Parkway Avenue, CN-600
Trenton, New Jersey 08625
Telephone: 609-530-2834
Fax: 609-530-3767

Brenda Ramanathan
Environmental Section, Room 213
New Mexico State Hwy & Trans. Dept.
P.O. Box 1149
Santa Fe, New Mexico 87504
Telephone: 505-827-0967
Fax: 505-827-6862

William McColl
Environmental Analysis Bureau
New York State Dept. of Transportation
State Campus, 5-303
Albany, New York 12232
Telephone: 518-457-2385
Fax: 518-457-6887

Stephen Walker
North Carolina Dept. of Transportation
Planning & Research Branch
P.O. Box 25201
Raleigh, North Carolina 27611
Telephone: 919-733-7844
Fax: 919-733-9794

Ben Kubischta
North Dakota Dept. of Transportation
600 E. Boulevard Avenue
Bismark, North Dakota 58505-0700
Telephone: 701-328-3555

Elvin W. Pinckney
Ohio Dept. of Transportation
25 South Front Street
Columbus, Ohio 43215
Telephone: 614-466-5154
Fax: 614-728-7368

Dawn Sullivan
Environmental Studies, Planning Division
Oklahoma Dept. of Transportation
200 Northeast 21st Street
Oklahoma City, Oklahoma 73105
Telephone: 405-521-2535
Fax: 405-521-6917

David Goodwin
Environmental Services
Oregon Dept. of Transportation
1158 Chemekcta Street NE
Salem, Oregon 97310
Telephone: 503-986-3488
Fax: 503-986-3524

Mark Lombard
Pennsylvania Dept. of Transportation
Forum Place, 7th Floor
555 Walnut Street
Harrisburg, Pennsylvania 17101-1900
Telephone: 717-772-2569
Fax: 717-772-0834

Irma M. Garcia
Puerto Rico Highway & Transportation Authority
P.O. Box 42007
San Juan, Puerto Rico 00940
Telephone: 787-729-1583
Fax: 787-727-5503

Ken Burke
Rhode Island Dept. of Transportation
Highway Engineering Division
2 Capitol Hill
Providence, Rhode Island 02903
Telephone: 401-222-2023
Fax: 401-222-3006

Wayne Hall
Environmental Section
South Carolina Dept. of Transportation
Post Office Box 191
Columbia, South Carolina 29201
Telephone: 803-737-1395
Fax: 803-737-9868

James D. Nelson
South Dakota Dept. of Transportation
700 E. Broadway Avenue
Pierre, South Dakota 57501-2586
Telephone: 605-773-3098
Fax: 605-773-6608

Raymond Brisson
Tennessee Dept. of Transportation
Environmental Planning Division
900 James K. Polk Building
Nashville, Tennessee 37243-0334
Telephone: 615-741-2612
Fax: 615-532-8451

Mike Shearer
State Department of Transportation
125 E. 11th D-8E
Austin, Texas 78071-2483
Telephone: 512-416-2622
Fax: 512-416-2319

Robb Edgar
Utah Dept. of Transportation
Materials Division
Box 148410
Salt Lake City, Utah 84114-8410
Telephone: 801-887-3402

Dennis Benjamin
Technical Services Division
Agency of Transportation
133 State Street
Montpelier, Vermont 05602
Telephone: 802-828-3978
Fax: 802-828-3983

Cary Adkins
Environmental Division
Virginia Dept. of Transportation
1401 East Broad Street
Richmond, Virginia 23219
Telephone: 804-371-6765
Fax: 804-786-7401

Martin Palmer
Washington State Dept. of Transportation
15700 Dayton Avenue North, MS-138
Seattle, Washington 98133-9710
Telephone: 360-440-4544
Fax: 360-705-6833

James Colby
West Virginia Dept. of Transportation
State Capitol Complex, Bldg. 5, Rm A-464
Charleston, West Virginia 25305
Telephone: 304-558-2885
Fax: 304-558-1334

Jay Waldschmidt
Wisconsin Dept. of Transportation
Bureau of Environment, Room 451
4802 Sheboygan Avenue, P.O. Box 7965
Madison, Wisconsin 53707-9806
Telephone: 608-267-9806
Fax: 608-266-7818

Timothy L. Stark
Environmental Services Engineer
Wyoming Dept. of Transportation
P.O. Box 1708
Cheyenne, Wyoming 82003-1708
Telephone: 307-777-4379
Fax: 307-777-4193

16.6 Training Courses

There are several training courses provided on the basics of highway traffic noise, including a special focus on highway traffic noise prediction and barrier design (from the acoustic standpoint). For more information on available training courses, contact Robert Armstrong at:

Federal Highway Administration, HEPN
400 Seventh Street
Washington, DC 20590
(202) 366-2073
(202) 366-3409
Robert.E.Armstrong@FHWA.DOT.GOV

REFERENCES

1. Simpson, Myles A. Noise Barrier Design Handbook. Report No. FHWA-RD-76-58. Arlington, VA: Bolt Beranek and Newman, Inc., February 1976.

2. Knauer, Harvey S., Soren Pedersen, Cynthia S.Y. Lee, Gregg G. Fleming, Out of the Box Productions. FHWA Highway Noise Barrier Design Video. Cambridge, MA: John A. Volpe National Transportation Systems Center, Acoustics Facility, February 2000.

3. Pedersen, Soren, Harvey S. Knauer, Cynthia S.Y. Lee, Gregg G. Fleming. FHWA Highway Noise Barrier Design CD-ROM. Cambridge, MA: John A. Volpe National Transportation Systems Center, Acoustics Facility, February 2000.

4. Anderson, Grant S., Cynthia S.Y. Lee, Gregg G. Fleming. FHWA Traffic Noise Model,® Version 1.0: User's Guide. Report No. FHWA-PD-96-009 and DOT-VNTSC-FHWA-98-1. Cambridge, MA: John A. Volpe National Transportation Systems Center, Acoustics Facility, January 1998.

5. Menge, Christopher W., Christopher J. Rossano, Grant S. Anderson, Christopher J. Bajdek. FHWA Traffic Noise Model,® Version 1.0: Technical Manual. Report No. FHWA-PD-96-010 and DOT-VNTSC-FHWA-98-2. Cambridge, MA: John A. Volpe National Transportation Systems Center, Acoustics Facility, February 1998.

6. Fleming, Gregg G., Amanda S. Rapoza, Cynthia S.Y. Lee. Development of National Reference Energy Mean Emission Levels for the FHWA Traffic Noise Model,® Version 1.0. Report No. FHWA-PD-96-008 and DOT-VNTSC-FHWA-96-2. Cambridge, MA: John A. Volpe National Transportation Systems Center, Acoustics Facility, November 1995.

7. Bowlby, William, Theodore Patrick, Cynthia S.Y. Lee, Gregg G. Fleming. FHWA Traffic Noise Model,® Version 1.0: Trainer CD-ROM. Cambridge, MA: John A. Volpe National Transportation Systems Center, Acoustics Facility, March 1998.

8. "Acoustical Terminology." American National Standard, ANSI S1.1-1994. New York: American National Standards Institute, 1994.

9. "Methods for Determination of Insertion Loss of Outdoor Noise Barriers." American National Standard, ANSI S12.8-1998. New York: American National Standards Institute, 1998.

10. "Procedures for Outdoor Measurement of Sound Pressure Level." American National Standard, ANSI S12.18-1994 New York: American National Standards Institute, 1994.

11. Johnson, D.L., A.H. Marsh, C.M. Harris. "Acoustical Measurement Instruments." Handbook of Acoustical Measurements and Noise Control. New York: Columbia University, 1991.

12. "Specification for Sound Level Meters." American National Standard, ANSI S1.4-1983 (R1997). New York: American National Standards Institute, 1997.

13. "Ground Impedance - Measurement of Ground Impedance." American National Standard. New York: American National Standards Institute, (to be published).

14. "Test Method for Laboratory Compaction Characteristics of Soil Using Standard Effort." American Society of Testing and Materials, ASTM Standard D 698-91(1998). Philadelphia, PA: American Society of Testing and Materials, 1998.

15. "Test Method for Sound Absorption and Sound Absorption Coefficients by the Reverberation Room Method." American Society of Testing and Materials, ASTM Standard C 423-90a. Philadelphia, PA: American Society of Testing and Materials, 1990.

16. "Classification for Rating Sound Insulation." American Society of Testing and Materials, ASTM Standard E 413-87 (R1994). Philadelphia, PA: American Society of Testing and Materials, 1987.

17. Highway Noise Barriers: Performance, Maintenance, and Safety (Video). Cambridge, MA: John A. Volpe National Transportation Systems Center, Acoustics Facility, October 1996.

18. Hendriks, Rudolf. Technical Noise Supplement - A Technical Supplement to the Traffic Noise Analysis Protocol. Sacramento, CA: California Department of Transportation, October 1998.

19. Lee, Cynthia S.Y., Gregg G. Fleming. Measurement of Highway-Related Noise. Report No. FHWA-PD-96-046 and DOT-VNTSC-FHWA-96-5. Cambridge, MA: John A. Volpe National Transportation Systems Center, Acoustics Facility, May 1996.

20. Embleton, T.F.W. "Sound Propagation Outdoors - Improved Prediction Schemes for the 80's." Noise Control Engineering. Vol. 18, No. 1, pp. 30-39, 1982.

21. Embleton, T.F.W., J.E. Piercy, and G.A. Daigle. "Effective Flow Resistivity of Ground Surfaces Determined by Acoustical Measurements." Journal of the Acoustical Society of America. Vol. 74, pp. 1239-1244, 1984.

22. "Method for Calculation of the Absorption of Sound By the Atmosphere." American National Standard, ANSI S1.26-1995. New York: American National Standards Institute, 1995.

23. "Acoustics - Attenuation of Sound During Propagation Outdoors - Part 1: Calculation of the Absorption of Sound By the Atmosphere." International Organization for Standardization, ISO 9613-1:1993. Geneva, Switzerland: International Organization for Standardization, 1993.

24. Hendriks, R.W. Field Evaluation of Acoustical Performance of Parallel Highway Noise Barriers Along Route 99 in Sacramento, California. Report No. FHWA/CA/TL-91/01. Sacramento, CA: California Department of Transportation, Division of New Technology, Materials and Research, January 1991.

25. Wayson, Roger L., Bowlby, William. "Atmospheric Effects of Traffic Noise Propagation." Transportation Research Board Annual Meeting, Paper No. 890796. Washington, DC: Transportation Research Board, January 1990.

26. Acoustics and Your Environment - The Basics of Sound and Highway Traffic Noise (Video). Cambridge, MA: John A. Volpe National Transportation Systems Center, Acoustics Facility, February 1999.

27. Kurze, U.J. "Prediction Methods for Road Traffic Noise." Lecture Notes: International Seminar of Road Traffic Noise Evaluation by Model Studies. Grenoble, France: September 1988.

28. Foss, Rene N. "Single Screen Noise Barrier." Noise Control Engineering. pp. 40-44, July-August 1978.

29. Barry, T.M. and J.A. Reagan. FHWA Highway Traffic Noise Prediction Model. Report No. FHWA-RD-77-108. Washington, DC: Federal Highway Administration, December 1978.

30. "Test Method for Impedance and Absorption of Acoustical Materials by the Impedance Tube Method." American Society of Testing and Materials, ASTM Standard C 384-95a. Philadelphia, PA: American Society of Testing and Materials, 1995.

31. Bradley, J.S., J.D. Quirt, J.A. Birta. "Measuring Sound Absorption of Traffic Noise Barriers." The Wall Journal, Issue 31. Lehigh Acres, FL: The Wall Journal, 1997.

32. Oostveen, J.C.J. "Gaps Beneath Noise Screens - Acoustically Permissible?" DWW wijzer, ISSN 0926-8618. Delft, Netherlands: Rijkswaterstaat, Road and Hydraulic Engineering Division, 1989.

33. Fleming, G.G. and E.J. Rickley. Performance Evaluation of Experimental Highway Noise Barriers. Report No. FHWA-RD-94-093. Cambridge, MA: John A. Volpe National Transportation Systems Center, Acoustics Facility, April 1994.

34. Fleming, G.G. and E.J. Rickley. Parallel Barrier Effectiveness: Dulles Noise Barrier Project. Report No. FHWA-RD-90-105. Cambridge, MA: John A. Volpe National Transportation Systems Center, Acoustics Facility, May 1990.

35. Hendriks, Rudolf. " To Absorb Or Not To Absorb." The Wall Journal. Issue 21, 1996.

36. Chalupnik, J.D. Multi-Level Roadway Noise Abatement. Report No. WA-RD-266.1. Seattle, WA: Washington State Transportation Center, University of Washington, April 1992.

37. Summary of Noise Barriers Constructed By December 31, 1995. Washington, DC: U.S. Department of Transportation, Federal Highway Administration, December 1996.

38. Cohn, Louis F. and Roswell A. Harris. Special Noise Barrier Applications: Literature Review - Task 1 (draft). Louisville, KY: University of Louisville, December 1991.

39. May, D.N. and M.M. Osman. "Highway Noise Barriers: New Shapes." Journal of Sound and Vibration. Vol.71, No.1, pp. 73-101, 1980.

40. Hutchins, D.A., J.W. Jones, and L.T. Russell. "Model Studies of Barrier Performance in the Presence of Ground Surfaces, Part II - Different Shapes." Journal of Acoustical Society of America. Vol. 75, No. 6, pp. 1817-1826, 1984.

41. Fujiwara, K. and N. Futura. "Sound Shielding Efficiency of a Barrier with a Cylinder at the Edge." Noise Control Engineering Journal. Vol. 37, No. 1, pp. 5-11, 1991.

42. Hutchins, D.A., J.W. Jones, B. Patterson, and L.T. Russell. "Studies of Parallel Barrier Performance by Acoustical Modeling." Journal of Acoustical Society of America. Vol. 77, pp. 536-546, 1985.

43. California Noise Barriers. Special Task Force on Noise Barriers, CALTRANS, June 1992.

44. "New-Product Evaluation Procedures." NCHRP Synthesis of Highway Practice, 90. Washington, D.C: Transportation Research Board, June 1982.

45. "Value Engineering in Preconstruction and Construction." NCHRP Synthesis of Highway Practice, 78. Washington, D.C: Transportation Research Board, 1981.

46. "Standard Method of Measuring Relative Resistance of Wall, Floor, and Roof Construction to Impact Loading." American Society of Testing and Materials, ASTM Standard E695-79(1997)e1. Philadelphia, PA: American Society of Testing and Materials, 1997.

47. "Standard Test Method for Surface Burning Characteristics of Building Materials." American Society of Testing and Materials, ASTM Standard E84-98e1. Philadelphia, PA: American Society of Testing and Materials, 1998.

48. "Methods of Test for Concrete." Canadian Standards Association, CAN/CSA Standard A23.2-5C. Etobicoke, Ontario: CSA International, June 1994.

49. "Air Content of Plastic Concrete." Canadian Standards Association, CAN/CSA Standard A23.2-4C. Etobicoke, Ontario: CSA International, 1994.

50. "Standard Test Method for Making, Accelerated Curing, and Testing Concrete Compression Test Specimens." American Society of Testing and Materials, ASTM Standard C684-96. Philadelphia, PA: American Society of Testing and Materials, 1996.

51. "Standard Test Method for Splitting Tensile Strength of Cylindrical Concrete Specimens." American Society of Testing and Materials, ASTM Standard C496-96. Philadelphia, PA: American Society of Testing and Materials, 1996.

52. "Standard Test Method for Microscopical Determination of Parameters of the Air-Void System in Hardened Concrete." American Society of Testing and Materials, ASTM Standard C457-90. Philadelphia, PA: American Society of Testing and Materials, 1990.

53. "Standard Test Method for Resistance of Concrete to Rapid Freezing and Thawing ." <u>American Society of Testing and Materials, ASTM Standard C666-97</u>. Philadelphia, PA: American Society of Testing and Materials, 1997.

54. "Standard Test Method for Scaling Resistance of Concrete Surfaces Exposed to Deicing Chemicals." <u>American Society of Testing and Materials, ASTM Standard C672-92</u>. Philadelphia, PA: American Society of Testing and Materials, 1992.

55. Janssen, D.J. and M.B. Snyder. "Resistance of Concrete to Freezing and Thawing." <u>Strategic Highway Research Program (SHRP-C-391)</u>. Washington, D.C: Strategic Highway Research Program, June 1994.

56. "Standard Practice for Operating Salt Spray (Fog) Apparatus." <u>American Society of Testing and Materials, ASTM Standard B117-97</u>. Philadelphia, PA: American Society of Testing and Materials, 1997.

57. "Standard Practice for Operating Light-Exposure Apparatus (Xenon-Arc Type) With and Without Water for Exposure of Nonmetallic Materials." <u>American Society of Testing and Materials, ASTM Standard G26-96</u>. Philadelphia, PA: American Society of Testing and Materials, 1996.

58. "Standard Test Method for Evaluating Degree of Checking of Exterior Paints." <u>American Society of Testing and Materials, ASTM Standard D660-93</u>. Philadelphia, PA: American Society of Testing and Materials, 1993.

59. "Standard Test Method for Evaluating Degree of Cracking of Exterior Paints." <u>American Society of Testing and Materials, ASTM Standard D661-93</u>. Philadelphia, PA: American Society of Testing and Materials, 1993.

60. "Standard Test Method for Evaluating Degree of Blistering of Paints." <u>American Society of Testing and Materials, ASTM Standard D714-87(1994)e1</u>. Philadelphia, PA: American Society of Testing and Materials, 1994.

61. "Standard Test Method for Calculation of Color Differences From Instrumentally Measured Color Coordinates." <u>American Society of Testing and Materials, ASTM Standard D2244-93</u>. Philadelphia, PA: American Society of Testing and Materials, 1993.

62. "Standard Test Methods for Measuring Adhesion by Tape Test." <u>American Society of Testing and Materials, ASTM Standard D3359-97</u>. Philadelphia, PA: American Society of Testing and Materials, 1997.

63. "Standard Test Methods for Evaluating the Degree of Chalking of Exterior Paint Films." <u>American Society of Testing and Materials, ASTM Standard D4214-98</u>. Philadelphia, PA: American Society of Testing and Materials, 1998.

64. "Standard Method for Sample Preparation for Determining Penetration of Preservatives in Wood." *American Wood-Preservers' Association, AWPA Standard A19-93*. Granbury, TX: American Wood-Preservers' Association, 1993.

65. "Sawn Timber - Determination of the Average Moisture Content of a Lot." *International Organization for Standardization, ISO 4470:1981*. Geneva, Switzerland: International Organization for Standardization, 1981.

66. "Standard for Noise Barriers on Roadways." *Canadian Standards Association, CAN/CSA Standard Z107.9 (draft)*. Etobicoke, Ontario: CSA International, November 1994.

67. "Test Method for Laboratory Compaction Characteristics of Soil Using Modified Effort." *American Society of Testing and Materials, ASTM Standard D1557-91(1998)*. Philadelphia, PA: American Society of Testing and Materials, 1998.

68. "Specification for Acoustical Calibrators." *American National Standard, ANSI S1.40-1984 (R1997)*. New York: American National Standards Institute, 1997.

69. "Electroacoustics - Sound Calibrators." *International Electrotechnical Commission, IEC 60942 Ed. 2.0 b:1997*. Switzerland: International Electrotechnical Commission, 1997.

70. "Measurement of Sound Pressure Levels." *American National Standard, ANSI S1.13-1995*. New York: American National Standards Institute, 1995.

71. "Electroacoustics - Instruments for Measurement of Aircraft Noise - Performance Requirements for Systems to Measure One-Third-Octave Band Sound Pressure Levels in Noise Certification of Transport-Category Aeroplanes." *International Electrotechnical Commission, IEC 61265 Ed. 1.0 b:1995*. Switzerland: International Electrotechnical Commission, 1995.

72. "Quantities and Procedures for Description and Measurement of Environmental Sound, Part 1." *American National Standard, ANSI S12.9-1988/Part 1(R 1993)*. New York: American National Standards Institute, 1993.

73. Bowlby, William, John Higgins, and Jerry Reagan. *Noise Barrier Cost Reduction Procedure - STAMINA 2.0/OPTIMA: User's Manual*. Report No. FHWA-DP-58-1. Washington, DC: Federal Highway Administration, April 1982.

74. Hendriks, R.W and Dick Wood. *Sound32*. Sacramento, CA: California Department of Transportation, Division of New Technology, Materials and Research, January 1991.

75. Bowlby, W., J. Li, and C. Patton. *TrafficNoiseCAD Version 2.0 -- User's Guide*. Brentwood, TN: Bowlby and Associates, Inc., April 1994.

76. Jones, Keith and Rudolf Hendriks. *Traffic Noise Analysis Protocol - For New Highway Construction and Reconstruction Projects*. Sacramento, CA: California Department of Transportation, October 1998.

77. The Sounds of Our Time (Video). Tallahassee, FL: Florida Department of Transportation, 1997.

78. Sound Advice (Video). Baton Rouge, LA: Louisiana Department of Transportation, December 1995.

79. Highway Traffic Noise - A Community Primer (Video). Trenton, NJ: New Jersey Department of Transportation, Bureaus of Environmental Analysis and Research, 1988.

80. Making Sound Decisions About Noise Abatement (Video). Harrisburg, PA: Pennsylvania Department of Transportation, 1996.

INDEX

-A-

A-weighted Sound Level ... 19
A-weighting ... 17, 19
Absorption Coefficient ... 9, 11, 25, 208
Absorption Material ... 42
Access ... 32, 41, 56, 57, 59, 62, 74, 96, 114, 116, 117, 133-137, 140, 145, 154, 163, 164, 169-171
Access Doors ... 114, 116, 134, 136, 140
Access Openings ... 57, 136, 145
Acoustic Effectiveness ... 187, 194, 195
Acoustic Energy ... 3-6, 11, 12
Acoustic Instrumentation ... 178, 179, 181-183, 189, 191, 192
Acoustical Considerations ... 1, 15, 27, 33, 41
Acoustics Facility ... 1, 207-209
Adjacent Land Uses ... 99, 109-111, 116
Aesthetic Considerations ... 104
Air Turbulence ... 20, 21, 166, 180
Alignment Changes ... 99, 100, 116
Aluminum ... 26, 42, 66, 67, 70, 91
Ambient Noise ... 3
American National Standards Institute (see also ANSI) ... 187, 198, 207, 208, 212
American Society of Testing and Materials (see also ASTM) ... 11, 208-212
Anchor System ... 50
Anchorage ... 39, 51, 52
ANSI S1.13 -971 ... 182, 191
ANSI S1.13-1995 ... 182, 191, 212
Anti-graffiti Coatings ... 93
ASTM B117 ... 151
ASTM C666 ... 150
ASTM C672 ... 150
ASTM D1557 ... 157
ASTM E695 ... 148
ASTM E84 ... 149
ASTM G26 ... 151
Atmospheric Absorption ... 20, 31, 35, 188
Atmospheric Conditions ... 188, 189
Atmospheric Effects ... 20, 21, 35, 180, 181
Attachments ... 38, 39, 50, 58, 60, 83, 84
Availability of Replacement Parts ... 42, 164, 170, 172

-B-

Background Noise ... 3
Barrier Absorption ... 22, 25
Barrier Design Goals ... 27, 35

Preceding Page Blank

Barrier Design Process 1, 10, 177, 185
Barrier Effectiveness 187, 209
Barrier End Treatments 103
Barrier Height Limitations 127
Barrier Insertion Loss 27, 28, 187-191, 193, 194, 197
Barrier Length 27-29, 35
Barrier Overlap Sections 133, 134
Barrier Sound Transmission (see also STC) 25, 35
Barrier Surface Treatment 80
Barrier Tops 34, 168
Berm 27, 29, 30, 35, 37, 38, 48, 49, 54, 59, 60, 111, 132, 161, 184, 194
Berming 103, 104
Bin Type Barriers 45
Brick 38, 44, 55, 64, 65, 83-86, 91, 93, 97, 123, 127, 131
Bridge Beams 169
Bridge Construction 51
Bridge Deck 10, 52
Bridge Parapet 52, 130
Bridge Slab 51
Bridge Structures 66
Burial Panel Type 46

-C-
Cap 57, 80, 102, 103, 106, 116, 136, 140, 168
Cast-in-place 160
Casting Process 149
Ceramic Tile 84
Characteristics of Sound 15
Climbability 67
Coatings 66, 68, 73, 76, 77, 79, 80, 91-95, 145, 159, 162, 165-167, 170
Color 42, 64, 70, 71, 81, 90-94, 96, 98, 105, 106, 109, 110, 151, 153, 158, 159, 163, 211
Combination Noise Berm and Noise Wall System 37, 48, 49, 60
Community Acceptance 1, 195
Community Noise Equivalent Level (see also L_{den}) 4, 8, 19
Community Participation 186
Composite 3, 77-79, 84, 89, 93, 130, 131, 144, 179
Composite Noise Barrier Materials 77
Compressive Strength 63, 65, 97, 149
Concrete 3-8, 10, 12, 13, 20, 26, 33, 38-41, 43-48, 50-52, 56, 57, 59-67, 72, 76, 77, 79, 81-85, 90, 91, 93, 94, 96-98, 119, 122, 126, 128, 130-132, 144, 148-151, 157, 158, 160, 166, 172, 210, 211
Concrete Footings 126, 128
Construction Noise 157, 160, 161
Continuous Footings 4, 62, 126
Corrosion Resistance 68, 97
Cost Considerations 1, 171

-D-

Day-night Average Sound Level (see also L_{dn}) 5, 8
Dead Load 4, 5, 42, 51, 52, 55, 59, 125, 127
Deicing Chemicals 65, 165, 211
Destructive Testing 158, 162
Dimension 9
Direct Burial Panels 38, 46, 47, 59
Divergence 6, 19, 20
Drainage 13, 27, 38, 47, 56-60, 96, 103, 114, 116-119, 122, 127, 130, 161
Drainage Openings 58
Drainage Requirements 59, 60, 116, 117
Dry-Cast Concrete Mix 64, 149

-E-

Early Construction of Permanent Noise Walls 161
Earthtone Colors 105
Emergency Access 57, 114, 133, 135, 136, 140, 164
Enamel 66, 91
End Treatments 103
Equivalent Sound Level (see also L_{AeqT}) 6, 8, 12, 19
Exposed Aggregate 41, 48, 60, 81, 84, 93, 96, 97, 106

-F-

Federal Highway Administration Traffic Noise Model (see also FHWA TNM) 189, 217
FHWA TNM 1, 194, 197
FHWA Traffic Noise Model 1, 184, 194, 197, 207
Fiberglass 74, 77, 92
Fir 26, 69, 70
Fire Hose Openings 114
Fire Hose Valves 114, 116
Fire Hoses 136, 137, 145
Fire Hydrant 136
Foam 3
Footing 3, 4, 13, 39, 42-45, 48, 56, 111, 119, 120, 122, 126, 127, 144, 154, 157
Footings 4, 13, 56, 62, 120, 122, 126, 128, 144, 145, 157
Form Liner 41, 61, 82, 89, 90, 96
Form work 48, 50, 60, 81, 82, 96, 149, 160, 172
Foundation 3, 7, 10, 33, 38, 39, 46-49, 56, 65, 67, 71, 97, 125-127, 157, 159, 160, 162
Foundation Attachments 38, 39
Foundation Requirements 125-127, 157
Fractured Fin 85
Free Standing Noise Walls 38, 44
Fresnel Number 7, 24, 29

-G-

Galvanized Finish	91
Galvanized Surfaces	159, 162
Galvanizing Material	66, 67
Gap	3, 21, 27, 32, 35, 43, 50, 117, 119, 140
Girts	7, 66, 67, 97
Glare	67, 73, 75, 97, 98, 138-140
Glass	26, 42, 72-74
Glass Panels	42, 72
Glue-laminated	70
Graffiti	54, 92-95, 106, 114, 145, 166, 167, 170
Graffiti Coatings	93, 94, 166, 167, 170
Groove Planking	70, 97
Ground Attenuation	8, 27, 187
Ground Impedance	8, 188, 192, 208
Guard Rail	113, 131, 132
Gunite	89

-H-

Hard Ground	8, 11, 20
Horizontal Caps	102, 116

-I-

Ice	7, 8, 61, 125, 127, 139, 145
Ice Credations	145
Impedance Tube Method	25, 209
Independent Foundation	56
Insect Damage	8
Insertion Loss	8, 9, 27, 28, 31, 34, 187-191, 193-195, 197, 207
Installation Considerations	1, 157, 160
Installation Defects	163
Installation Jigs	160
Instrumentation	6, 177-183, 189, 191, 192
Iron	51, 66, 81, 96
Iron Ore	66, 81, 96

-L-

L_{AE} (see also Sound Exposure Level)	4-6, 8, 12, 19
L_{AaeqT} (see also Equivalent Sound Level)	6, 8, 12
L_{AFmx} (see also Maximum Sound Level)	9, 19
L_{ASmx} (see also Maximum Sound Level)	9, 19
L_{den} (see also Community Noise Equivalent Level)	4, 8, 19
L_{dn} (see also Day-night Average Sound Level)	5, 8, 19
L_{10} (see also Ten-percentile Exceeded Sound Level)	8, 13, 19
Labor Costs	174

Laminated Panels 87
Landscaping 56, 57, 59, 62, 96, 109, 111-116, 154, 165, 167, 170
Lighting 13, 54, 58, 60, 72, 117, 119, 122, 133
Line Source 6, 7, 9, 19, 20
Line-of-sight 21, 27
Liner 41, 61, 82, 89, 90, 96
Liner Finish 41, 61, 89
Liner Pattern 82, 96
Litter 59, 116, 167, 170
Litter Removal 116, 167, 170
Live Load 9
Load Factor 9
Lumber 13, 68, 71, 152

-M-

Maintenance Considerations 1, 41, 42, 54, 114, 163
Masonry Block 38, 44, 52, 64, 65, 84, 85, 91, 97, 131
Maximum Sound Level (see also L_{AFmx} and L_{ASmx}) 9, 19
Metal . 4, 6, 7, 9, 10, 40, 56, 57, 66-68, 70, 82, 86, 89, 91, 97, 125, 130-132, 134, 138, 139, 148, 150, 151, 158,
159, 161, 165, 166, 179
Meteorological 19, 20, 177, 180, 182, 187-189, 191, 192
Meteorological Conditions 182, 188, 191, 192
Meteorological Data 180
Meteorological Effects 19
Microphone 177-183, 188-194

-N-

Noise Barrier Aesthetics 1, 99
Noise Barrier Materials 1, 61, 77, 165
Noise Barrier Types 1, 37
Noise Berm 37, 48, 49, 59, 60, 161
Noise Reduction Coefficient (see also NRC) 9, 25, 76, 77, 79, 80, 98, 145
Noise Wall 1, 37, 38, 43-50, 52-61, 80, 87, 99, 102, 103, 113, 116, 120, 122, 132, 134-136, 161, 166
Noise Walls Used to Partially Retain Earth 38, 47, 59
Nondestructive Test 71, 152
Non-destructive Testing 162
NRC (see also Noise Reduction Coefficient) 9, 25, 76, 77, 79, 80
Overlapping Barriers 32, 33, 35, 133

-P-

Parallel Barrier 10, 31, 32, 197, 209, 210
Parapet 10, 50-52, 60, 125, 130
Pattern 41, 59, 82, 87, 90, 96, 102, 106, 107, 110
Pedestrian Access Openings 57
Petroleum Products 76, 79

219

Pigmented 83, 90-94, 96-98, 166
Pine 69, 77, 80, 149
Plantings 37, 45, 59, 67, 103, 112, 114, 116, 132, 168
Plastic 6, 42, 45, 46, 57, 72-75, 78, 88, 92, 98, 139, 140, 149, 210
Plywoods 68
Point Loads 10, 42
Point Source 6, 7, 10, 19
Precast Concrete 38, 43-46, 57, 59, 62, 63, 81, 119, 122
Pressure Treating 70, 71, 152
Product Evaluation Process 1, 146, 147

-R-

Recyclability 77, 79
Recycled Rubber 76, 88, 92, 98
Reference Microphone 181-183, 188, 190-194
Reflective Barrier 30
Reflective Panels 138
Reflective Parallel Barriers 31
Reflective Surface 8, 20
Refraction 20, 21, 35
Repairs 71, 75, 78, 124, 163, 164, 166, 169
Retaining Wall 11, 13, 48, 49, 54-57, 124, 169
Reverberation Room Method 25, 208
Road 32, 73, 103, 105-108, 116, 130-132, 165, 168, 201, 209
Rubber 76, 77, 79, 82, 88, 92, 98
Rubber Materials 88
Rusting 67, 97, 103, 165

-S-

Sabine Absorption Coefficients 9, 25
Safety 1, 2, 26, 27, 32, 45, 53, 56-60, 69, 72, 74, 76-79, 96-98, 116, 122, 129, 131-133, 136, 138, 140, 143,
                                                            148, 163, 165, 166, 168, 198, 208
Safety Considerations 1, 58, 129
Salt 14, 53, 61, 63, 64, 68, 76, 79, 80, 93, 150, 151, 165, 205, 211
Sampling Period 177, 181, 182, 189, 191
Shape 33, 34, 52, 62, 63, 81, 83, 105, 108, 109, 111, 116, 131, 132, 150, 157
Shatter Resistance 69, 73, 75, 78, 79, 98, 139, 140, 148
Sight Distance 53, 60, 72, 74, 132, 140
Sign Supports 117, 119, 122
Site Grading 157, 162
Slump Test 63, 149
Snow Drifting 54, 60, 168, 170
Snow Loads 125
Snow Storage 168
Soft Ground 7, 11, 20

FHWA Highway Noise Barrier Design Handbook — *References*

Sound Absorption Coefficient 11
Sound Exposure Level (see also L_{AE}) 4-6, 8, 12, 19
Sound Pressure Level 3, 6, 12, 13, 15, 20, 25, 178, 207
Sound Transmission Class 12, 27, 61, 67, 76, 77, 79, 80, 97, 98, 145
Spalling 12, 64, 84, 150, 151, 158
Spread Footings 13, 62, 126
Stains 94, 95, 158, 159, 166, 167, 170
STC (see also Sound Transmission Coefficient) 12, 27, 67, 76, 77, 79, 80
Steel 14, 26, 39, 40, 42, 48, 62, 66, 67, 70, 72, 91, 92, 97, 98, 103, 113, 131, 144, 145, 158, 164, 172
Stone Crib 7, 38, 44, 46, 47, 59
Storage of Materials 157, 159
Structural Considerations 1, 42, 55, 123
Structure-mounted Noise Walls 49
Stucco 84, 85, 89, 96, 97
Surface Treatment 41, 61, 80, 82, 89, 150, 165, 170
Swales 95, 118

-T-

Temporary Noise Walls 161
Ten-percentile Exceeded Sound Level (see also L_{10}) 8, 13, 19
Textures 41, 48, 80, 84, 86, 88, 89, 96, 97, 105, 106, 109, 110, 116
Thnadner 34
Tilted 43, 59, 73, 139, 140
Timber 7, 71, 212
TNM 1, 1, 184, 194, 197, 198
Tongue and Groove Planking 70, 97
Toxicity 76, 77, 80, 95, 98, 149
Traffic Barriers 62, 69, 79
Traffic Protection 132, 133, 173, 175
Transmission Loss 12, 13, 25, 26, 35
Transparent Materials 88
Transparent Panels 72, 98, 139

-U-

Ultraviolet Light 14, 68, 73, 75, 77, 78, 80, 93, 98, 166, 170
Urethane 94, 95
Utilities 13, 59, 117, 120-122, 125, 127, 140, 154, 169, 173
Utility Considerations 1, 117

-V-

Vehicular Impact 53, 129, 131, 132, 139
Veneers 84, 96, 97
Vertical Caps 103

-W-

Warping 26, 40, 70, 71, 74, 78, 92, 97, 98, 103, 152, 158, 166
Weather Conditions 21, 48, 80, 151
Weatherometer Testing 14, 77, 80
Width-to-height 31
Wind Load 43, 125
Wood ... 4, 5, 7, 8, 10, 12, 13, 26, 39, 40, 43, 45, 46, 68-71, 77, 82, 86, 87, 92, 97, 98, 130-132, 134, 144, 148, 152, 158, 166, 212
Wood Posts 39, 40, 71, 87
Wood Preservative Treatments 92, 98
Wood Species 92, 98

www.ingramcontent.com/pod-product-compliance
Lightning Source LLC
Chambersburg PA
CBHW080239180526
45167CB00006B/2343